Building Identities: Socio-Political Implications of Ancient Maya City Plans

Matthew S. Mosher

BAR International Series 2111
2010

Published in 2016 by
BAR Publishing, Oxford

BAR International Series 2111

Building Identities: Socio-Political Implications of Ancient Maya City Plans

ISBN 978 1 4073 0652 0

© MS Mosher and the Publisher 2010

The author's moral rights under the 1988 UK Copyright,
Designs and Patents Act are hereby expressly asserted.

All rights reserved. No part of this work may be copied, reproduced, stored,
sold, distributed, scanned, saved in any form of digital format or transmitted
in any form digitally, without the written permission of the Publisher.

BAR Publishing is the trading name of British Archaeological Reports (Oxford) Ltd.
British Archaeological Reports was first incorporated in 1974 to publish the BAR
Series, International and British. In 1992 Hadrian Books Ltd became part of the BAR
group. This volume was originally published by Archaeopress in conjunction with
British Archaeological Reports (Oxford) Ltd / Hadrian Books Ltd, the Series principal
publisher, in 2010. This present volume is published by BAR Publishing, 2016.

Printed in England

BAR titles are available from:

	BAR Publishing
	122 Banbury Rd, Oxford, OX2 7BP, UK
EMAIL	info@barpublishing.com
PHONE	+44 (0)1865 310431
FAX	+44 (0)1865 316916
	www.barpublishing.com

ABSTRACT

BUILDING IDENTITIES: SOCIO-POLITICAL IMPLICATIONS OF ANCIENT MAYA CITY PLANS

Questions pertaining to city-planning and political organisation have long intrigued students of Maya archaeology. For the most part, these issues have been pursued separately, usually from the stance of either architectural analysis or epigraphy. However, recent observations on a possible connection between public architecture and the display of inter-regional political power have led some scholars to posit that a non-epigraphic means to assess ancient Maya political hierarchies and alliances may be through civic-plan analysis. This thesis attempts such an exercise in examining the evidence for ancient Maya "superstates" as gleaned from an analysis of ancient Maya city plans. While the evidence does imply a correlation between inter-and intra-regional power blocs and shared city plan elements, it also importantly suggests that such correlations may have as much to do with common Maya understandings and expressions of cosmology and ritual integration as with polity relations.

ACKNOWLEDGEMENTS

This thesis would have been impossible to undertake without the time and efforts of many different individuals in many different capacities. Drs. John Topic and Paul Healy served as members of my supervisory committee and left lasting impressions on me from my time at Trent, not the least of which was their willingness to give of their time, experience, and materials in order that I better understand and appreciate the discipline in which I had found myself. Dr. Arlen Chase acted as the external reviewer of my thesis and I thank him for his participation, comments, and encouragement. Dr. Wendy Ashmore, on whose work this thesis is based, has my sincere gratitude for providing me with unpublished material and encouraging a test of her ideas on the link between ancient Maya architecture and politics. Dr. Gyles Iannone was my thesis supervisor, and is responsible for introducing me to fieldwork in the Maya area. His guidance and persistent encouragement were invaluable.

Also I would like to thank the staff at the Department of Anthropology at Trent University, especially Janice Ecclestone and Kerri-Anne Pierce. Their assistance through my time at Trent was invaluable.

I did more than study anthropology at Trent, I also made some good friends. The following are thanked for their inspiration, for listening to me moan on and on about ancient Maya politics, and for genuine friendship: Julie Kapyrka, Katie Bittner, Ann Albers and Nathan Content, Tom Dorman, Sarah MacGregor, Ryan Primrose, Dan Lambert, James E. Herbert, José Martinez and Miriam Khalil, and David Rewniak. Additionally, the following are thanked for many long hours spent discussing the finer points of my thesis: Meaghan Perumaki-Brown, Dr. Jeffrey D. Seibert, Adam Pollock, Jöelle Chartrand, Barbara Slim de Proulx.

It is an understatement to say that without the support of my family, both immediate and extended, this project would never have been possible. For logistical, financial and, above all, emotional support I would like to thank the following: Marc Lepage and Ashleigh Mosher, Kiera and Mercedes Lepage, Kyleigh, Adam, Greyden and Piper Campbell, Revs. B. Alex and Debra Mosher, Garnet Parker, Murray and Sandra Hines, Eugene and Shirley Hines, Rick and Sylvia Didsbury, Les and Marion Didsbury.

Since the completion of my M.A. thesis, I have branched into the study of other geographic areas and cultural traditions, in large part because of the more encompassing questions relating to political organization, urbanism, and the built environment this study brought to my awareness. In this process I have learned a great deal from many people, and they, in turn, have been forced to endure my ignorance and comparisons with Classic Maya data. They are Drs. Heather M.-L. Miller and Edward R. Swenson, Jennifer Campbell, Martin T. Bale, and Greg V. Braun. I thank them all.

Above all, I would like to thank my wife, Amber, for her love, support, and understanding. She has taken care of everything in my life in order that I may periodically leave her for months at a time and dig in foreign dirt. To her, as well as my son Daegan, and daughter Rowan, I owe everything.

TABLE OF CONTENTS

Abstract and Acknowledgements	1
List of Tables	3
List of Figures	3
Preface and Disclaimer	4
Chapter 1: Introduction	6
Research Objectives	6
Chapter 2: Background Information	9
Political Models	9
Decentralised States	9
Segmentary States	10
Galactic Polities	10
City-States	11
Centralised States	11
The Dynamic Model	13
Superstates	13
Tikal and Calakmul in Maya History	13
Civic Plans and Political Organisation	14
Cosmology	15
The Multilayered Universe	15
The Axis Mundi	16
Directionality	16
Maya Cities and Cosmology	17
Summary	18
Chapter 3: Theoretical Considerations	19
Landscape	19
The Built Environment	20
City Planning	21
Mesoamerican Civic Aesthetics	21
Architecture as Communication	24
Summary	25
Chapter 4: Methods and Data Presentation	26
Architectural Markers of Tikal and Calakmul	26
Tikal Paradigmatic	26
Tikal Syntagmatic	26
Calakmul Paradigmatic	27
Calakmul Syntagmatic	27
Problems with the Methodology	27
Political Structure	28
Site Maps/Analysis	28
Paradigmatic Elements	28
Syntagmatic Elements	29
Summary	30
Chapter 5: Interpretation	31
Common Elements	31
Politically Specific Elements	31
Calakmul Paradigmatic Element 1	32

Calakmul Syntagmatic Element 1	32
Calakmul Syntagmatic Element 4	32
Tikal Syntagmatic Element 3	32
Ancient Ritual Integration: A Possible Explanation	32
City-State Culture	33
Summary	34
Chapter 6: Conclusions	35
Summary	36
Appendices	69
Bibliography	70

LIST OF TABLES

Table 1:	Primary Political Affiliates of Calakmul and Tikal	37
Table 2:	Secondary Political Affiliates of Calakmul and Tikal	37
Table 3:	Ritual Activity Amongst Polities	38

LIST OF FIGURES

Figure 1: Map of Tikal, Guatemala	39
Figure 2: Map of Calakmul, Mexico	40
Figure 3: Map of Aguateca, Guatemala	41
Figure 4: Map of Arroyo de Piedra, Guatemala	42
Figure 5: Map of Bonompak, Mexico	43
Figure 6: Map of Cahal Pichik, Belize	44
Figure 7: Map of El Cayo, Mexico	44
Figure 8: Map of Cancuen, Guatemala	45
Figure 9: Map of Caracol, Belize	46
Figure 10: Map of Comalcalco, Mexico	47
Figure 11: Map of Dos Pilas, Guatemala	48
Figure 12: Map of Holmul, Guatemala	49
Figure 13: Map of Ixkun, Guatemala	50
Figure 14: Map of Jimbal, Guatemala	51
Figure 15: Map of La Pasadita, Guatemala	52
Figure 16: Map of Tamarandito, Guatemala	52
Figure 17: Map of La Muñeca, Mexico	53
Figure 18: Map of Nadzcaan, Mexico	54
Figure 19: Map of Naranjo, Guatemala	55
Figure 20: Map of Nim Li Punit, Belize	56
Figure 21: Map of Palenque, Mexico	57
Figure 22: Map of Piedras Negras, Guatemala	58
Figure 23: Map of Pomona, Mexico	59
Figure 24: Mapped Sections of Pusilha, Belize	60
Figure 25: Map of Quirigia, Guatemala	61
Figure 26: Map of Tonina, Mexico	62
Figure 27: Map of Ucanal, Guatemala	63
Figure 28: Map of Xunantunich, Belize	64
Figure 29: Map of Yaxchilan, Mexico	65
Figure 30: Map of Uaxactun, Guatemala	66
Figure 31: Map of Copan, Honduras	67
Figure 32: Map of Yaxha, Guatemala	68

PREFACE AND DISCLAIMER

What follows is an unmodified version of my MA anthropology thesis completed at Trent University in 2004 (for illustrative purposes certain maps have been substituted with later versions). Obviously, many changes have taken place within Maya archaeology since that time, not least of which the degree to which studies of urbanism and urban planning feature as a regular component of archaeological inquiry (Smith 2007, 2008, 2009). This is in keeping with a growing appreciation of and proclivity for dialogue and collaboration between students of complex societies from different geographical and disciplinarian traditions (Bagwell 2006; Blanton and Fargher 2008; Chrisomalis 2006; Cowgill 2004; Marcus 2009; Marcus and Sabloff 2008; Smith 2007, 2008, 2009; Smith and Schreiber 2005, 2006; Trigger 2003; Yoffee 2005). As a result, new data and synthetic theoretical frameworks have greatly changed the way Mesoamerican scholars approach and interpret urban planning (Smith 2007, 2008). My own thinking on the subject has indeed changed, and if I were to conduct this study again, I doubt I would do so in an identical manner (cf. Mosher and Jones 2008). Three issues in particular should be read with this caveat in mind: the theoretically structuralist methods I utilize to distinguish the particular architectural characteristics of the Tikal and Calakmul-influenced civic layouts, the discussion of Classic Maya political systems, and my somewhat religiously deterministic explanation of Classic Maya civic space.

Methods

In this volume I present a methodological approach derived from the work of Ashmore (1986, 1989, 1991, 1992) and Houk (2003) and ultimately based upon a semiotic approach to architectural analysis (Hodder 1987b). I suggest that Classic Maya civic layouts are composed of (and are in fact reducible to) both syntagmatic and paradigmatic elements. That is, common architectural elements (ballcourts, temple-pyramids and the like) can be "read" to communicate specific cultural information. This takes two forms: syntagmatic and paradigmatic communication. Syntagmatic meaning is derived from the specific ordering or grouping of communicative elements, resulting in a syntactical, or "surface" level of meaning (Yule 1996:100-103). This is complemented by paradigmatic meaning, which alludes to the deeper structure which lies behind a syntagmatic understanding. As such, I advocate that certain groupings of Classic Maya architectural elements, when examined in conjunction, betray a "deeper" paradigmatic meaning which I relate to the long-standing political divisions of the Late Classic. While I still think this approach is a useful one, I do not necessarily clearly explicate the differences between synagmatic and paradigmatic elements beyond the introduction of the concepts in Chapter 3, which may lead to some obfuscation between the terms in the chapters which follow.

Political Organization

In attempting to situate and make sense of the discussions concerning Classic Maya political systems I present a comparison of "centralized", and "decentralized" states and touch upon segmatary states, galactic polities, and superstates. I would do so differently now. A state is the political expression of a civilization (Baines and Yoffee 1998, 2000). It is by definition centralized, being predicated on the monopoly of political agency by an elite ruling class (Hansen 2000:13;). Although differences exist in the particular strategies and historical conventions between different states, to speak of "decentralised" (or "weak") states creates a false category which obscures the commonalities inherent in this form of political organization (Smith 2004). Although I elude to it in the discussion on political organization, my distinction between "centralized" and "decentralized" states should have been cast as a part of the larger discussion which served to distinguish city-states from their territorial counterparts (Charlton and Nichols 1997; Hansen 2000). On a superficial level, city-states may indeed appear to exhibit less political control when compared to territorial ones; however, this is an issue of the characteristics and strategies of governance inherent in the these types of polities, rather than an absolute measure of the grasp of the ruling elite (Trigger 2003:113-117). True, direct government or administrative involvement in various aspects of a society may shift over time within a particular polity (hence the true utility of the Dynamic Model (Marcus 1993, 1998)), but the structural foundation of statehood as a centralized authority remains. Similarly, the term "galactic polity" (Demarest 1992, 2004:212-217), relating to Geertz's (1980) theatre state concept which emphasizes the primacy of ideological investment as the foundation of political authority, should have been discussed as an aspect or particular strategy of statehood rather than an independent form of state. All states, regardless of temporality or location, are in part held together and justified through the mechanisms of political ideologies (Bawden 1995; Baines and Yoffee 1998, 2000; Yoffee 2005:38-40). The Classic Maya were no exception, but neither were they solely reliant on the elite manipulation of religious and political ideology implied by the galactic polity model, as the growing evidence for the importance of the economic and administrative facets of Classic Maya statehood make clear (Dahlin et al 2007; Inomata 2001; Pyburn 2008; Scarborough and Valdez 2009). Also under the heading of decentralized states, is the term "segmentary states." Although well entrenched within studies of Classic Maya political organization, this term should have been avoided altogether. As Marcus (1998:8) makes exceptionally clear, this term refers to fissioning within segmentary lineage systems, a key means of social organization amongst *non-state* societies. As with the discussion of galactic polities, my treatment of segmentary states was misconceived, and should have appeared as a discussion on the ideas concerning the importance of elite factionalism as a facet of Classic Maya states (Lucero 2007). Lastly, the term "superstates" may be thought of as hegemonic systems of allied polities dominated by Tikal and Calakmul, respectively. Such affiliations are not uncommon occurrences within city-state civilizations (Hansen 2000:17) and do not represent a separate stage of political evolution.

Civic Space

The Classic Maya, like many ancient peoples, conflated religious and political authority, a reality which was expressed though the proliferation of cosmological imagery imbued in various state-sponsored or elite media, including architecture. However, this state of affairs does not dictate that the Classic Maya gave primacy to religious or cosmological concepts over more temporal ones. Contrary to the language employed in this study, the strategic use of directional imagery and religious precepts in civic planning does not constitute the use of cosmograms in such activity (Smith 2005). As Smith and others have pointed out: a "cosmogram" implies replicating the structure of the cosmos, the creation of an architectural or other model of how a particular people conceives of the universe. Rather, the Classic Maya selectively used certain key cosmological concepts and directional symbols, especially those which were amenable to articulating royal authority with the religious realm of Classic Maya life, such as the east-west solar circuit which, properly portrayed, quite nicely provided a potent metaphor for the regenerative, eternal, and natural character of Classic Maya kingship (Freidel 2008). Also, despite the relative wealth of Classic period texts, no epigraphic evidence explicitly links civic space with the emulation of cosmological ideals. This has not dissuaded scholars of the Classic Maya from utilizing language which assumes such a connection, a practice of which I am guilty throughout the following pages.

CHAPTER 1

INTRODUCTION

This thesis examines, through a variety of evidence, Late Classic (c. 250-900 AD) Maya political organization, specifically the existence of large-scale political structures as evidenced through specific patterns of city plans and architectural similarities. This particular exercise draws upon such interconnected aspects of current and past Maya scholarship as epigraphic reconstructions of political history, elite architecture, the nature of the ancient Maya state, and research into the less tangible aspects of the ancient Maya civilization, such as the cosmological and ideological frameworks within which such issues were conceived, negotiated, and imbued with meaning.

Traditionally, academic interest in ancient Maya political society has been characterized by two opposing interpretations: 1) Maya polities were either autonomous city-states exercising authority over their immediate hinterland, or 2) such polities were largely regulated and controlled by a few main political centres, effectively creating large, regional states (Adams 1981; Martin and Grube 1994, 1995, 2000; Mathews 1991). The evidence cited for both of these views has likewise been dichotomized; on the one hand, scholars have focused on monumental architecture and issues of urbanism and settlement patterns, and, on the other, an attempt to wrest historical meaning from Maya hieroglyphic texts. Often, these two perspectives have resulted in a divergence rather than a convergence of ideas about the overall character of ancient Maya society; studies of architecture and urbanism have been used to generate models of socio-political complexity, stressing the co-evolution of social hierarchy and monumentality, while the decipherment of the ancient Maya writing system, having begun in earnest only within the last generation, has revealed much in the way of site- specific political dynasties and their social relationships to dynasties at other sites (Martin and Grube 2000; Schele and Mathews 1991). We have thus crafted interpretive models which emphasise the rise and intensification of localized socio-political complexity but which lack political dynamism, at the same time as we have revealed historical details of political agency without reference to the physical settings in which such activities occurred.

Recently, however, scholars have attempted to remedy this apparent disunion through the integration of text and archaeology, but often without adequate attention to the role of architecture as a meaningful expression of political identity (Ashmore 2002; Houston 1998). Rather, city plans simply remain the backdrop against which intra- and inter-polity activity is carried on. This approach runs counter to recent developments in landscape archaeology theory which stresses the experiential aspects of landscapes, including the intentionality (political and otherwise) of the built environment as well as its communicative role (Andrews 1975; Ashmore and Knapp 1999; Barrett 1999a; Bender 1992; Kowalski 1999; Reese-Taylor 2002; Reese-Taylor and Koontz 2001).

Following these developments, Wendy Ashmore and Jeremy Sabloff (2002) have recently claimed that the layout of the Classic Lowland Maya polity centres may encode information pertinent to the study of ancient Maya political organisation. They propose that there may be a connection between political affiliations and the architectural arrangement of Classic Lowland cities. In particular, they suggest that many, if not most, Classic Lowland cities arranged their epicentre architecture after the fashion of Tikal, Guatemala, or Calakmul, México. This is significant precisely because these are the polities upon which the notion of ancient Maya superstates is based. Briefly, the "superstate" model, advocated by Simon Martin and Nikolai Grube (1995, 2000), argues that rather than regional states, the ancient Maya political system was characterised by hegemonic city-states, a situation where autonomous city-states were tied up in political blocs dominated by the most militarily and politically powerful polities: Calakmul and Tikal. The epigraphic record abounds with references to the royal lineages of Tikal and Calakmul, their political manoeuvrings, expansionism, and longstanding mutual rivalry.[1] The physical emulation of either of these polities' epicentre designs then may have political implications.

Alternately, the resemblance of city centres to one another may imply broader, more subtle social processes. As "participants" in historical events and the physical expression of political and religious ritual, architecture itself (here implying civic layout) serves to symbolise, and thus, communicate patterns of cultural experience (Leach 1976). As Reese-Taylor and Koontz (2001) explain through the application of cultural poetics and aesthetics, notions of what is "right" are culturally constituted and historically contingent. Perhaps what we are witnessing in the similarity of Maya city layouts is adherence to shared cultural views of political and social legitimacy as expressed through architecture, rather than explicit statements of political affiliations. This leaves open the interpretation of agency in the layout of ancient Maya cities, and forces us to think critically about just what emulation is, how it operates, and what it entails.

In order to address these contending ideas, it is necessary to review the pertinent political histories of Late Classic Maya polities in conjunction with the architectural evidence for city designs. At the same time, we must acknowledge that our understanding of both of these

---

[1] Although I subscribe to the notion that Tikal's main political and military rival during the Late Classic period was indeed Calakmul, not all scholars agree. Epigraphers have long known of Tikal's adversarial relationship with an ancient Maya polity known solely through hieroglyphic texts, a polity commonly referred to as "Site Q." Epigraphic and archaeological research by Marcus (1973, 1976) first suggested that Calakmul and Site Q were indeed the same locale. However, since Marcus' initial proposal, new data have put Calakmul's status as Site Q into question (Chase and Chase 1998:20-21).

areas is handicapped due to an incomplete material record.

RESEARCH OBJECTIVES

During the Classic period (c.250-900 AD) of ancient Maya civilisation, the lowland regions of Guatemala, the Yucatan Peninsula, Mexico, northwest Honduras, western El Salvador, and the entirety of Belize were host to numerous urban centres, each usually headed by separate, local royal dynasties which participated, through kin ties and a shared ethos of nobility, in a pan-Maya elite culture. These political elite generally made use of a common set of architectural features through which they performed administrative and religious functions, and in which they resided. These features included range structures (what is popularly referred to as the Maya "palace"), ball courts, astronomical assemblages, and pyramidal terraces topped with temples (Andrews 1975; Houston 1998; Loten and Pendergast 1984). By far the largest and most architecturally complex of such centres in the Maya lowlands were Tikal and Calakmul (Adams and Jones 1981; Martin and Grube 2000). Aside from their imposing size, these cities were the seats of the most prominent Classic period polities, each enjoying far-reaching military and political influence. Written records attest to their mutual antagonism, which seems to have persisted for the majority of the Classic period. Furthermore, the position of most dominant lowland polity seems to have alternated almost exclusively between them (Martin and Grube 1994, 2000; Schele and Freidel 1990).

These cities' opposition in matters political and military appears to be reflected in their civic arrangement. The layout of Tikal's epicentre (Figure 1) seems to be based on an elaborated version of a twin pyramid group (Ashmore 1991). The twin pyramid group complexes, almost exclusive to Tikal, are specific arrangements of architecture comprised of two pyramidal structures facing one another along an east-west axis and separated by a plaza. This plaza is bounded on the north by a structure iconographically related to the veneration of royal individuals, and on the south by an equally symbolic structure related to ancient Maya concepts of the underworld (Ashmore 1986, 1989, 1991, 1992; Jones 1969).

Conversely, the centre of Calakmul (Figure 2) is notable for the elongated north-south orientation of its principal plaza bisected by a pyramidal structure. Both the north and south extremities of this plaza feature elite architecture – in the north, a (presumed) residential/administrative structure, and in the south the colossal Structure 2, a pyramidal acropolis of Preclassic origins (Folan 1992; Folan et al. 1995a, 2001). Significantly, the east and west of the central plaza are framed by an E-Group assemblage (an architectural arrangement which features a western situated pyramidal structure facing three, usually connected, eastern structures). Both the Tikal and Calakmul epicentre patterns are assumed to have evoked distinct references from a common corpus of ancient Maya (and indeed broader Mesoamerican) ideology and cosmological symbolism (Ashmore 2002; Reese-Taylor and Koontz 2001; Schele and Kappelman 2001).

Following on the common assumption that amongst the ancient Maya, like most archaic civilisations, the realms of religion and governance were thoroughly intertwined, these different epicentre "templates" would have been imbued with both cosmological and political significance (Ashmore and Sabloff 2002; Trigger 2003:79-88). The later adoption of either of these epicentre templates by other ancient Maya polities may, therefore, denote socio-political affiliation with either Calakmul or Tikal, perhaps as participants in, or subjects of, a hegemonic superstate. Alternately, such behaviour may indicate the appropriation of established authoritarian or prestige symbols, or simply the generic manner in which the ancient Maya conceived of the socially proper method of organising their urban space.

In order to clarify these issues and determine just what, if anything, the physical, architectural arrangements of Calakmul and Tikal meant to those who lived in these ancient cities, I address the following questions:

1. What are the specific architectural factors which define the Tikal and Calakmul layouts?
2. What specific cosmological, ideological, and experiential aspects of ancient Maya society are alluded to in these templates?
3. Does the emulation of these templates imply appropriation of political and religious ideals or membership in a superstate?
4. What was the operational level of emulation? Were all the polities considered in this study actively involved in fashioning their respective centres after Tikal or Calakmul, or was there a more dispersed pattern, where secondary and tertiary centres modelled themselves after allied primary centres which aimed to reproduce the seat of the Tikal and Calakmul polities?
5. Does the civic plan evidence correlate with the epigraphically reconstructed political history?

Such questions follow current practice within Mesoamerican archaeology, as the study of site orientation and layout is receiving increasing attention (Houk 2003). Many such studies are explicitly concerned with the relation of architecture and site designs to cosmology and how this relates to overall socio- political organisation. Monte Albán, San Lorenzo, and, most extensively, Teotihuacán, have been subject to recent growing inquiries into these very issues (Cowgill 2000; Flannery and Marcus 1999; Orr 2001; Sugiyama 1993). Within the Maya sub-area of Mesoamerica, it has been argued that the physical manipulation of hieroglyphs within elite architecture can create a politicised space, such as demonstrated at Yaxchilán, México (McAnany and Plank 2001). Likewise, ideology and secular authority converge in the placement of elite ritual structures at Palenque, Mexico, and Dos Pilas, Guatemala (Brady 1997; Brady and Ashmore 1999). Where I diverge from these examples is in seeking to place such symbolic behaviour within the context of a larger socio-political structure, and not to address issues of ideology of place as something that was site-specific.

Surely, as the ideological dimensions of architecture have been illustrated to play an important role in the historic processes of individual polities, could not these same notions have had wider implications, such as a public means of articulating relations of power between such polities? This is in accord with the current ideas on the purposeful construction of "imagined communities" amongst the ancient Maya elite (Isbell 2000; Yaegar 2000).

As symbols of belonging, ancient Maya city plans may have been similar in function to other markers of elite power (such as hieroglyphic texts (Marcus 1992)), but much more accessible as resident populations "experienced" them on a corporeal basis (Barrett 1999b; Bender 1992; Tilley 1994). This is an important point and one which ties the present study to larger issues in archaeological theory, such as phenomenology and praxis. However, we cannot lose sight of the fact that "experience" is, and always has been, a constant process of negotiation between (possibly conflicting) social, temporal, and physical realities (Fletcher 1992, Hodder 1987a, 1999; Iannone 2002). In short, meaning is contextual, and the study of ancient Maya city plans may have more inherent interpretive value when approached from this perspective than from assigning political conditions based on the presence or absence of material traits (Hodder 1987b, 1999). The present study, by examining both the syntagmatic and paradigmatic meanings found within ancient Maya city layouts, hopes to offer a refined and encompassing statement on such symbols.

In order to achieve such goals the present work is divided by the following chapter headings: Background Research, Theoretical Considerations, Methods and Data Presentation, Interpretation, and Conclusions.

The Background Research section (Chapter 2) presents in detail the conceptual foundations of this thesis –Ashmore and Sabloff's (2002) proposal which links ancient Maya city plans and political alliances, and the "superstate" model as proposed by Martin and Grube (1995, 2000). As a matter of reference and clarification, significant space is devoted to the competing models of political organisation which have been utilised to understand the ancient Maya state. Likewise, the current state of knowledge on ancient Maya cosmology is summarised, in particular Ashmore's (1989, 1991) work relating to the cosmological significance of city plans.

Chapter 3 addresses the theoretical context of the exercise, and places it within a growing body of scholarship inquiring into the connection between culture and landscape. A brief review of settlement studies –the primary approach to landscape amongst Mayanists – is provided, as are models for understanding civic designs. The main emphasis of the latter is addressed through discussions of cultural aesthetics, semiology, and praxis.

The Methods and Data Presentation section of the thesis (Chapter 4) presents the manner in which I organised ancient Maya political alliances and in which I discerned the analytically diagnostic components of ancient Maya civic plans. In regard to the latter, the recognition of syntagmatic and paradigmatic "chains of evidence" for civic plans is the means of analysis.

Chapter 5 (Interpretation), presents the results of the data analysis, and posits that other, more subtle integrative social mechanisms than overt political jockeying may in fact account for what similarities may exist in ancient Maya civic plans. The documented uses of political and religious ritual as a form of elite interaction by the Tikal and Calakmul ruling factions throughout the Late Classic period figure prominently in this discussion.

Chapter 6 offers a summary overview of the research questions presented at the outset of this exercise, and highlights where future research could, and, in my opinion, should, expound on the themes discussed herein.

CHAPTER 2

BACKGROUND RESEARCH

Ancient Maya political organisation and cosmology have proven to be among the more enduring preoccupations of epigraphers and archaeologists. Throughout the course of Maya studies these issues have impacted upon each other, and are today largely studied as interconnected facets of a complex elite ideology (Demarest 1992; Fash 2002; McAnany and Plank 2001; Taube 1993).

Conceptualising ancient Maya understandings of politics and cosmology has also proven to be one of the most frustrating of Mayanists' tasks. Partly, this is due to the fact that these are encompassing subjects; studies by Marcus (1976, 1993), McAnany (1995), Schele and Freidel (1990), and Schele and Miller (1986), suggest that cosmological underpinnings affected nearly every facet of ancient Maya society, from the institution of divine kingship to subsistence methods and kinship organisation. Likewise, political organisation bears directly on issues of economics, expressive media, and our understanding of ancient historical records (Culbert 1991; Fox et al. 1996; Marcus 1992; Sanders and Webster 1988). In the integrated study of cosmology and political organisation, then, we have to try to accommodate those components of ancient Maya society which, traditionally, have been treated as disparate issues. While this entails a retreat from strict academic specialisation, it offers a more holistic view of ancient Maya society, one envisioned by Willey (in an altered form) over two decades earlier (Willey 1980).

The following discussions will deal explicitly with aspects of ancient Maya cosmology, elite ideologies, and, to a lesser extent, religion. As a matter of clarification, and to avoid the ambiguity which often seems to accompany these issues within archaeology, I offer the following definitions:
> cosmology can be defined as a theory or philosophy of the origin and general structure of the universe, its components, elements, and laws, especially those relating to such variables as space, time, and causality. How the universe is structured affects both religion and ideology [Flannery and Marcus 1999:351-352].

Likewise, religion
> can be defined as a specific set of beliefs in a supernatural power or powers, to be obeyed and worshipped as the creator(s) and/or ruler(s) of the universe. [This includes a] worldview which lies at the interface of religion and cosmology [Flannery and Marcus 1999:353].

In addition to the above is the notion of ideology, which
> Can be defined as a body of doctrine, myth, and symbolism of a social movement, institution, class or group of individuals, often with reference to some political or cultural plan, along with the strategies of putting the doctrine into operation [Flannery and Marcus 1998:353].

Although I am primarily concerned with architectural expressions of ancient Maya cosmological ideals and political institutions, these topics necessitate a general overview of the scholarship into these larger issues. It should also be noted that I will refer to ancient Maya polity seats explicitly as cities, rather than ceremonial/civic/or administrative centres, sites, or other common terms. In part, this is to focus and define the scope of inquiry presented in this thesis; centres and sites are rather vague terms and can refer to a range of settlement units. Moreover, there exists consensus among modern scholars that in the ancient Maya we have an example of a truly urban civilisation, especially according to Mogens Herman Hansen's (2000) concept of "city-state culture" (Chase et al. 1990; Grube 2000; Haviland 1970; Marcus 1983; Trigger 2003).

POLITICAL MODELS

By virtue of their physical complexity and sheer size, the remains of the ancient Maya cities which so impressed Stephens and Catherwood in the mid 1800s inspired a century-long period of inquiry and debate into the type of society to which such creations could belong. Although the finer details continue to be worked out, scholars now concede that the ancient Maya society met all of the requisites to qualify as a state-level civilisation (Chase et al. 1990; Culbert 1991; Demarest 1992; Feinman and Marcus 1998; Iannone 2002; Martin and Grube 1994, 2000; Trigger 2003).

In general, there have been two broad models applied to understand the ancient Maya state: 1) the decentralised state (usually equated with city-states), and, 2) the view that ancient Maya polities were administrative, centralised units (often identified as regional states). Both of these approaches are characterised by internal divisions which makes discussing the ancient Maya state more complicated than addressing it as an either/or situation. It should also be noted that in addition to the dominant paradigms, a third interpretive framework attempts to typify ancient Maya polities as fluctuating between decentralised and centralised types of political organisation, across both space and time (the "Dynamic Model") (Iannone 2002; Marcus 1993, 1998). In order to better understand the role of architecture and cosmology in power relations amongst ancient Maya states, we must first examine the differing ramifications of these competing political models.

Decentralised States

The school of thought that has held the most favour among scholars regarding the character of the ancient Maya state is that which advocates a decentralised form of government (Ball and Tascheck 1991, 2001; Demarest 1992; Fox 1977; Fash 1988; Sanders and Webster 1988). In general, the decentralised view places ancient Maya states as inherently less burdened by administration and bureaucracy as means of control and governance, and within a system in which kinship ties (especially amongst the nobility) and the reciprocal obligations inherent in such structures, formed the ultimate basis of political authority. These assertions are based on many interconnected lines of evidence, both archaeological and ethnographic. In short, the main features of the decentralised state apparatus are as follows (after Iannone 2002):

1) Economic, administrative, and ritual redundancy. These characteristics are manifested most visibly in architectural repetition, i.e., the presence of royal residential complexes in numerous settings within a polity, which may indicate the duplication of respective offices. This also implies that issues of politics and administration, as well as ritual were highly localised, with each city tending to the needs of its own population and, hence, not subject to larger networks of such business.
2) Closely tied to the above issue is the notion of "provincial" capitals, of which all other communities in a state system are physical and functional emulative reproductions politically bound to the former. This is often interpreted as the underlying reason behind a two or three tiered settlement hierarchy.
3) Kinship as the primary (and, in some cases, only) means of social integration.
4) The manipulation of ideology (especially that relating to kinship or lineage membership as outlined above) as a means of establishing a "ritual hierarchy".
5) The tendency for misfortunes and providence (economic, military, etc.) to remain city-specific; thus reinforcing the insular nature of a weakly-integrated system of polities.
6) Insufficient integration to allow for more than minimal allegiance and involvement of peripheral communities. That is, those cities and populations which have the least vested interest in the state are those physically and socially peripheral to the capital, or seat of power. In turn, these are the factions which are most likely to shift political alliances between neighbouring states.

Segmentary States. The decentralised state construct most commonly applied to the ancient Maya is the segmentary state (derived from the work of ethnologist Adian Southall [1956]), which is characterised as heavily reliant on the familial (kinship) ties and the personal character of the ruling dynasty. The emphasis on vertical authority reaffirms the kin-like basis of political legitimacy which stresses the ultimate commonality of the populace. That is, the relation of the ruling house to the commons was analogous to the relations between a lineage founder and his descendants). Although all ancient states were composed of at least two (usually endogamous) social strata (the ruling and the ruled), segmentary-state advocates envision a central role for supra-strata lineage groupings (a situation where inclusive lineages cross such boundaries) (McAnany 1995:111-156; Trigger 2003:146).

Ethnographic and ethnohistoric accounts have been key inspirations for scholars working within this framework (Fox and Cook 1996). It is important, then, that the historic Yucatan Maya, known to us through the writings of Frey Diego de Landa (Tozzer 1941), conceived of their political capitals as those places in which the royal dynasty resided (Marcus 1993:128).

As the foremost of the nobility, the royal family constituted a focal point for local populations, providing an elaborated version of kinship authority characterised by the Maya extended household. These "house societies" are thought to have provided the economic, political, and ideological base upon which Maya states flourished (Iannone 2001, 2002; Lévi-Strauss 1962; Schele and Mathews 1998). A further point of support for this notion is that ancient polities were known by the toponym of their royal seats, thus extending the pattern of upward emulation from a local, and, hence, familial social structure (Grube 2000; Marcus 1993:128).

Owing to the heavy reliance on kin and lineage-based authority in segmentary states, scholars generally infer a relatively weak monarchical control over people and resources (Iannone 2002). The ability of ruling factions to extract corvée labour and agricultural surplus from the resident populations are seen as the realistic limits of royal power. Other social institutions, such as corporate landowning groups, temple or religious estates, and various cross-cutting societal organisations (such as moieties or military associations), were functionally separate from the state apparatus. As Trigger (2003:317-318) illustrates with the example of the Yoruba of Western Africa, such non-governmental institutions not only looked to their own affairs with a fair degree of autonomy, they were often in a more advantageous position, economically and politically, than were the ruling families themselves. Of course, it must be stated that within segmentary states, there often were close ties, kin-based and otherwise, between the ruling families and the heads of particular social orders (Trigger 2003:264-275). Nevertheless, as decentralists argue, it was the need of the ruling class to work with and through such groups that made political control within segmentary states inherently tenuous. Although an authoritarian structure based on kinship principles offered one avenue to maintain political control, the ruling stratum also secured its position through the incorporation (or appropriation) of shared ideological and cosmological values (Demarest 1992; Freidel 1983, 1992; Freidel and Schele 1988). This claim has given rise to a related version of the segmentary state known as the galactic polity.

Galactic polities. The galactic polity or theatre state model, promoted for the ancient Maya by Arthur Demarest (1992) following the earlier works of Clifford Geertz (1980) and Stanley Tambiah (1977), places a heavy investment in elite ideology and associated ritual as the primary means of social integration. This was manifested in the society's investiture in the office of divine kingship which, according to this model, was the symbolic glue that held together the disparate factions inherent in a segmentary system. The office of the king and the validation of his divine providence required continual reification through displays of public rituals, such as dance and feasting (Looper 1995; Miller 1998; Reents- Budet 2003). Involvement in such rituals was especially pertinent in forging and maintaining the links which bound separate cities and other settlement units together within a polity (Reents-Budet 2003; Yaegar 2000).

The galactic polity model is the decentralist position par excellence in that administrative issues (agricultural, economic, military, etc.) are not only seen as secondary social integrators, but actually represent spheres of interest peripheral to, and often beyond, the ultimate mechanics of social cohesion. Demarest (1992:146) states that "even if large-scale intensive agricultural systems were present in some areas, it does not follow that state control existed." Likewise, issues of political economy are only considered insofar as they relate to the exchange and conspicuous consumption of ideologically-charged elite materials (Demarest 1992:143).

One attribute of ancient Maya cities which has been used to support the existence of a galactic polity structure is the purposeful arrangement of civic architecture to construct giant cosmograms (Demarest 1992:153). As the word implies, a cosmogram is a diagram, or template, of the universe made physical through the strategic use of architecture and other landscape components as metaphors for cosmological tenets (Ashmore 1989; Carrasco 1990). Although the presence of cosmograms in ancient Maya and, indeed, Mesoamerican civic plans has been recognised by scholars who do not necessarily adhere to galactic polity models, it is the ritual intentions inherent in such meaningful arrangements that, according to this model, have the most interpretative value for reconstructing socio-political integration (Ashmore 1991; Cowgill 2000; Demarest 1992:153; Reese-Taylor 2002; Trigger 2003:468). According to the advocates of this model, the cosmologically charged settings reinforce the ideological basis of the social order, and provide an integral, public component for elite ritual (Miller 1998).

City-states. Often, the decentralised models employed to make sense of ancient Maya socio-political organisation advocate that the ultimate unit of political authority was the city-state (Grube 2000). Typically, it is the "redundant" features listed above which lead to such an interpretation. The duplication of architectural forms (implying functional redundancy, and hence, non-specialty) and the public displays of shared ideological components of ancient Maya society (including the ubiquitous royal title k'uhul ahau [divine lord]) which reoccur throughout the Lowland polities, attest that it was the city-state (usually the polity capital and associated hinterland settlements) which characterised the highest level of ancient Maya political organisation (Martin and Grube 2000:18-19; see also Hansen 2000). This interpretation envisions a political landscape populated by functional equals, although, in reality, some "equals" came to have greater power and influence than others.

One aspect of the archaeological record which lends support to such an interpretation is the absence of any clearly-defined administrative or redistributive centres such as those that typify regional states like that of the Inca (Trigger 2003). However, it should not be assumed that decentralised states and city-states are interchangeable concepts. Some scholars who advocate a city-state model for understanding the ancient Maya state do so with the understanding that such units were highly centralised, like regional states, but effectively operated at a much smaller scale (Chase and Chase 1996a).

Centralised States
At the other end of the spectrum from decentralised socio-political organisation is the argument for the presence of centralised states among the ancient Maya. Centralised states, variously termed regional and territorial, assert a much stronger role for economics and institutionalized decision making as integrative devices (Hansen 2003; Trigger 2003). Rather than communicating the common ideology of Maya lineage structures in an elaborated "house society", the architectural dimensions of Maya cities essentially result from centralized planning and deliberate use of shared understandings of cosmology as a means of securing power (Fox et al. 1996:796; Iannone 2002). Dentralised states are, therefore, typically construed of as "organic", in the Durkheimian sense, and offer a more tightly integrated form of society than do the more "mechanical", decentralised states (Iannone 2002:69; Marcus 1993:113). That is, where decentralised states are socially integrated through a common investment in a shared identity, centralised states integrate populations through mutual specialisation and need (economic, agricultural, etc.).

Like their decentralised counterparts, centralised states are recognisable through a compilation of key traits (after Iannone 2002):

1. The presence of hierarchies of differentiation. This applies to economics, politics, administration, ritual, and all manner of things typically considered systemic subsets (Flannery 1968). In contrast to the decentralised case for the apparent "redundancy" of these same traits, this issue is often approached through architectural analysis which focuses on specialisation. Arguments by Chase and Chase (1996) assert that at Caracol there existed administrative and religious structures which were qualitatively different from those found in nearby dependencies, such as Cahal Pichik. This suggests one of two situations:
1) the architecture correlated with supervisory offices which administered certain aspects of the local dependencies; or 2) these architectural features provided services only available at Caracol proper, to which the dependant populations would have to travel. In either circumstance, what we have at Caracol is an example of centrally-placed architectural specialisation.

2. An emphasis on the separation between the rulers and the ruled. It was argued previously that under a

decentralised socio-political structure, integration was achieved, in part, through the promotion of a common identity. However, the reverse is expected for centralised states. Haviland (1992) has suggested that although lineages may have been ritually important, they were nevertheless downplayed in favour of more administrative means of social solidarity. An important means of underscoring the separateness of the elite and commoner strata was the use of architecture, especially restricted access courtyards and acropolis groups (Iannone 2001:127; Seibert 2002). This does not, however, imply that a common political identity was downplayed in centralised states (Chase and Chase 1996b).

Additionally, elite personal adornment and dress, rather than being solely accessible to the wealthy members of society, may have been restricted to the upper stratum. Although seldom explored in the literature, we should not discount the ethnographic analogies available from Central Mexico, an area which has many similarities with the Maya sub-area of Mesoamerica including, presumably, sumptuary laws (Berdan 1982).

3. State control and sponsorship of large-scale public works. In addition to elite and public ritual architecture being under royal control, centralists argue that water management, intensive agricultural features (such as hillside terracing and raised fields), and large scale defensive works would have necessitated an organisational complexity beyond the capabilities of a decentralised state (Chase and Chase 2001a). Hence, the presence of such features at places like Caracol
(Healy et al. 1983), Minanha' (Primrose 2003, 2004; Pollock 2003), and Calakmul (Folan et al. 1995b) has been interpreted as secondary evidence of an administrative "middle class", working on behalf of a central political authority.

4. State control of information. Within ancient states, the official control of information is most clearly seen in the state's exclusive (or de facto) right to use and display written language (Marcus 1992; Trigger 2003:143, 607-610).

Amongst the ancient Maya certain iconographic imagery may have also been restricted to use by the royal court (i.e., the "Jester God" [Freidel and Schele 1988b]).

5. Territorial units which were created and maintained through non-kinship means. This is perhaps the defining issue of centralised socio-political integration, and one in which the differences between this and decentralised models are summed up. As illustrated by McAnany (1995:95-108,125-130) for the decentralist model, political agency was largely bound to the ability of place-specific lineages to support and participate in the office of kingship. Therefore, there existed an innate connection between land and lineage. This is articulated through the use of the Yucatecan term ah kuch kabob, which simultaneously refers to land and "burden", a synonym for political authority and lineage obligation (McAnany 1995:92).

Within the hieroglyphic records, however, there exist numerous mentions of ahauob relations directly pertinent to territorial organisation, couched in hierarchical, rather than familial, language. This is most obvious in the expression y-ahau, or "his (subservient) lord", a reference to a political inferior (Martin and Grube 1994:9). The y-ahau phrase is often used in conjunction with another phrase, u-kahiy, "it was done by him" to commemorate the accession of subservient kings and validate their political role (Martin and Grube 1994:8).

Therefore, it would seem that the ultimate authority over territory is vested in the supervisory figure, one often removed geographically, and, hence, lineage-wise, from the territory in question. A well known example of this is the installation of K'ak Tiliw Chan Chaak (sometimes referred to as "Smoking Squirrel") of Naranjo by Yich'aak K'ak of Calakmul in 693 AD (Martin and Grube 2000:75).

The need of a royal figurehead to be conferred in his office by an outside political power suggests that the control and administration of territorial units were far removed from the hands of the lineages themselves (Chase and Chase 1996a; Grube 2000).

Additionally, and more applicable to archaeological inquiry, are the related issues of polity and population size. Chase and Chase (1996), Folan et al. (1995a), and Haviland (1969) have argued that the ancient Maya polities of Tikal, Caracol, and Calakmul enjoyed resident populations ranging from 50, 000 to over 100,000 individuals, encased within a minimum of 800 km². By any estimates (and especially those drawn from Maya ethnography [Haviland 1966; Vogt 1970]), these figures are far too large to be accommodated through lineage means. Such a situation argues for a regional state model more in accord with what we know historically of administrative states (Trigger 2003).

In general, the advocates for centralised political organisation among the Classic Maya, relying more on archaeological inference than ethnographic analogies to formulate their ideas (although by no means exclusively), argue against their decentralised counterparts because they feel that the physical evidence is indicative of a level of socio-political organisation far too complex to be accommodated within a decentralised framework. This argument, in addition to the points raised above, also takes issue with the concepts and terminology utilised by decentralists, the segmentary state model in particular (Chase and Chase 1996a; Feinman and Marcus 1998). As Chase and Chase (1996:803) state, "the basic component of the segmentary state is the segmentary lineage", something which has been critiqued as both inapplicable to the ancient Maya (although see Fox 1987, 1989 for more modern correlates), and as nothing other than a heuristic device. The centralists argue, therefore, that the decentralists are misguided in their interpretation of ancient Maya socio-political organisation in that they are trying to fit their data into ideal classificatory types (Chase and Chase 1996a:803). However, the same argument has been levied against the centralist position by advocates of the "Dynamic Model" (Iannone 2002).

The Dynamic Model
The third and most recent of the general models applied to the ancient Maya state is that dubbed the "Dynamic Model" (Iannone 2002; Marcus 1993, 1998). This model is, in part, a critique of the previous frameworks in that they do not explicitly acknowledge or address variation and change across time and space. Also, in part, this model accommodates principles from both decentralist and centralist positions. The dynamic model claims that decentralising and centralising processes were simultaneously present in ancient Maya polities, and that polities vacillated between the two positions over time.

The key architect behind the application of the dynamic model to the ancient Maya is Joyce Marcus (1993, 1998). Although she draws primarily upon ethnohistoric material to formulate and present her ideas, she claims that the processes of socio-political unification (centralisation), and dissolution (decentralisation) were active through the entire timespan of the ancient Maya state. Furthermore, she emphasises that both the epigraphic and archaeological records comply with the ethnohistoric examples of socio-political consolidation and balkanisation. For example, the coincidence of large-scale public construction and prolific hieroglyphic works declaring political and military expansion at Tikal contrast with periods of minimal construction and little, if any, production of textual records (Jones 1991). This can be taken as evidence of the "cycling" between periods of socio-political centralisation (expansion) and decentralisation (retraction). This same general situation also holds true (though less intensively) at other investigated ancient Maya cities, such as Calakmul (Folan et al. 1995a), and Piedras Negras (Houston et al. 2000).

The dynamic model has been investigated from several angles, including the search for centralising and decentralising forces (McAnany 1995) and structural history (Iannone 2002). One common denominator which is concurrent with non-Maya investigations into state-level socio-politics is the "fundamental tension between the institutions of kinship and kingship" (Iannone 2002:68; see also Trigger 2003:71-91). These areas of ancient Maya society, although connected (see above), nevertheless presented different opportunities, obligations, ideologies, and social arrangements for the respective populations. Different socio-political and historical conditions would have facilitated more or less social investment in one or the other of these competing institutions, resulting in centralised or decentralised polities, a typical pattern among ancient states (Feinman and Marcus 1998; Haviland 1977; Marcus 1998). This also suggests that social complexity is flexible and adaptive, rather than a confining, unidirectional, and indeed, evolutionary pattern (Haviland 1997; Iannone 2002:70).

More so than either the centralist or decentralist position, that of the dynamist reminds us that the models we employ to understand ancient states are heuristic devices, and, for an honest appreciation of the workings of such states, we must realize that flexibility and social agency are the norms.

Superstates
In addition to the particular form (or forms) of integration which characterised ancient Maya states, scholars have recently been focusing on the nature of past Maya political networks and intrapolity cohesion. In particular, researchers of historical texts have been leading the foray into this quagmire of ancient Maya political culture (Culbert 1988; Martin and Grube 1994, 1995, 2000). As with much of Maya scholarship, however, the exercise of unravelling political intricacies from the hieroglyphic record has raised as many questions as it has resolved. For example, using the same set of data, scholars have reconstructed a Late Classic political landscape populated by dozens of autonomous city-states, four large regional states, hegemonic power blocs, and territorial peer-polity confederacies (Adams and Jones 1981; Freidel 1986; Marcus 1973; Martin and Grube 2000; Mathews 1991; Sabloff 1986; Schele and Freidel 1990). As is painfully clear to students of ancient Maya socio-politics, the growing consensus on the meanings of the hieroglyphic record has been matched by a growing disagreement as to interpretive models.

Debates over the size and character of intrapolity units are nearly as old as Maya scholarship, and intimately bound to the issues raised in the state models discussed above. As Fox et al. (1996:795) note, academic opinion from the time of Morley (1935) has run the gamut of proposed models. However, it was not until the mid twentieth century decipherments by Proskouriakoff (1960) and Berlin (1958) that the proposed models of ancient Maya society began to be treated to the same level of scholarly contemplation as other early states.

One such proposal is that dubbed the "superstate" model of political organisation, as advocated by Grube and Martin (1994, 1995, 2000). This model follows the premise that, during the Late Classic period in particular, the majority of the Maya city states were politically aligned with either Tikal or Calakmul. This argument rests predominately on evidence from epigraphy which, having at its core the concerns and histories of the political elite, has revealed that the Tikal and Calakmul ahauob were the "overkings" of a number of other Maya cities.

Tikal and Calakmul in Maya History. Calakmul and Tikal, located in Southern Campeche, Mexico, and the Northern Peten, Guatemala, respectively, have long been considered dominant powers in their respective locales

(Marcus 1976, 1987; Schele and Freidel 1990). Initially, this distinction was owed to the impressive architectural sprawl of these two cities, as well as their voluminous hieroglyphic texts and monuments (Adams and Jones 1981; Folan et al. 1995a; Martin and Grube 2000). Known as the Mutul (Tikal) and Kaan (Calakmul) kingdoms in antiquity (Schele and Freidel 1990; Martin and Grube 2000:30,101), the recent decipherment of their written histories has revealed the existence of prolonged, and oftentimes severe, competition between the two cities.

As early as 546 AD, Calakmul had begun a program of "overseeing" the ruling houses of various central lowland cities –thus adding to its already established political role in relation to its northern neighbours (Martin and Grube 2000:103).

In 556 AD, Sky Witness, the reigning ahau of Calakmul, had gained the loyalty of the ruling house of Caracol, Belize. Previous to this, Tikal had counted Caracol among its own subject cities; this was in accord with Tikal foreign policy since at least 378 AD, when Uaxactun was fell under the political dominion of this larger neighbour (Braswell 2003; Schele and Freidel 1990:130-164).

The incidence of Caracol's wayward allegiance, in particular, seems to have set the tone for a conflict-ridden Late Classic period. The same year as Caracol's defection to Calakmul, Tikal launched an "axe event" against Caracol, usually thought to imply a military raid, and possibly concerned with ritual sacrifice (Martin and Grube 2000:39). In 562 AD, Caracol, together with Calakmul, declared a "star war" against Tikal –a type of religiously sanctioned conquest warfare "coinciding with a stationary node in the motion of Venus" (Martin and Grube 2000:39). Conceivably, the 130 year epigraphic silence which began at Tikal at this point was the result of such an action.

For the next century and a half Calakmul would hold a dominant position in the central lowlands. Many of Tikal's political allies were attacked directly by Calakmul (such as Palenque in 599 AD), as was Tikal itself (Martin and Grube 2000:42). While no evidence exists for overt Calakmul control of the Tikal polity, Calakmul was intensively involved in Tikal's domestic affairs through its support and endorsement of the breakaway Mutul state of Dos Pilas (Martin and Grube 2000:164; Fahsen 2003).

Through marriage ties, ritual activity, explicit statements of hierarchy and royal visits, scholars have revealed a pattern suggesting that the mutual antagonism between Tikal and Calakmul expanded in the Late Classic to form the basis of supra-polity power blocs, a "macro-political" form of organisation (Martin and Grube 1994; Peuramaki-Brown 2002; Schele 1991).

Although the exact nature of such superstates has yet to be agreed upon, I concur with Grube (2000) in his claim that these structures have all the markings of a hegemonic city-state system. Firstly, there does not appear to be any explicit appropriation of city-state territories. With the notable (yet brief) instance of the Calakmul/Caracol annexation of the Naranjo royal court in the 7th century, the polities included in the superstates seemed to have remained largely autonomous, self-governing units (Martin and Grube 2000; Mathews 1991). This may have to do with the ideological charter of ancient Maya kingship (Freidel 1992). The persistence of the ahau title, even as the subservient y-ahau, suggests that, fundamentally, Maya rulers claimed the same cosmological basis of power and authority. Therefore, the ritual and political co-option of royal figures, facilitated in part through joint public rituals, as claimed in Demarest (1992) (but also through more mundane means such as the threat of warfare (Webster 1993)), would have been key in cementing the dominion of Tikal and Calakmul over their peers (Grube 2000:550).

Secondly, the reconstructed territories of each superstate are non-contiguous (Martin and Grube 1995). This situation does not fit with the current academic definition of regional or territorial states (Trigger 2003).

Thirdly, there were hugely disparate differences in the size, populations, and political and military prominence between the focal points of superstates (that is, Tikal and Calakmul) and their constituents (Martin and Grube 1994, 2000). This negates the possibility of a "peer-polity" system of socio-political checks and balances (Freidel 1986; Grube 2000:550; Renfrew and Cherry 1986; Sabloff 1986).

Fourthly, and perhaps most telling of the nature of ancient Maya superstates, there is no instance in the epigraphic record of Late Classic polities being simultaneously aligned to Calakmul and Tikal (Martin and Grube 2000).

Therefore, it would seem that adherence to one of these superstates precluded an independent stance on external political relations. In short, we have a system of hegemonic polities and political dependencies (Grube 2000:550).

Civic Plans and Political Organisation. Although the presence of ancient superstates is far from settled within Maya epigraphy, the recent research of Ashmore and Sabloff (2002, 2003) at Xunantunich, Belize, and Sayil, Mexico, offers the potential of bringing it fully into the archaeological realm. In their analysis of these two cities, they conclude that the plans of both largely resemble those of the neighbouring cities – the larger, and presumably politically dominant, city Naranjo, Guatemala, and the smaller, and presumably politically dependant city Labná, México, respectively. Rather than ascribing such similarities to the random accretion which characterises a component of the growth cycle of any urban centre, the authors address the "ideational foundations" of architectural arrangement and urban landscape amongst the ancient Maya (Ashmore and Sabloff 2002:202).

The authors' discussion allow that such factors are inextricable from ancient Maya cosmological and political concerns, issues with which academics have had difficulty approaching as distinct categories of ancient Maya social experience (Ashmore 1989; Macri 1994; Walker and Lucero 2000). With the support of relevant developmental chronologies they arrive at the "provisional" conclusion that Xunantunich and Labná replicated the architectural metaphors that articulated power, both religious and political, at the larger centres (Ashmore and Sabloff 2002:202). Furthermore, the spatial similarities between Xunantunich and Naranjo are also present at Calakmul, arguably the most influential and powerful Late Classic polity in the Maya lowlands (Ashmore and Sabloff 2002:206-207). Such a powerful presence seemed a likely source of civic patterning for Naranjo itself. This indicated to the authors, then, that a broader process of purposeful civic arrangement was an active and meaningful element of ancient Maya urban design.

Additional research by Ashmore (1991, 2002) indicates that the plans of other ancient Maya cities in the central and southern lowlands share in the Calakmul layout, as well as in the civic pattern which she identifies as characteristic to Tikal (Ashmore 1986, 1989). The implications of such findings are that we have in the study of elements of such city plan elements a chance to explore the topic of political alliances and "superstates" through non-epigraphic means.

In order to address this, we must first consider another factor in ancient Maya architecture, and, indeed, politics; the issues surrounding the ancient Maya understandings of religion and the universe.

COSMOLOGY
From the infancy of Maya studies until the latter part of the last century, ancient Maya cosmology, religion and, to a large part, political structure were believed to have been concerned solely with calendrical and astronomical functions (Morley 1935). This was because, for the most part, our understanding of ancient Maya historical records was limited to matters mathematical; the extant Postclassic codices deal with calendrical ritual, and similarities between these texts and the Classic Period records bred an informed denial of the historical nature of the texts (Coe 1992). Although temporal and celestial matters do form an important component of ancient (and modern) Maya cosmology, it is acknowledged that equating such concerns with the entirety of said cosmology is unsatisfactory, and that it denies the role of the Maya creation mythology, and the metaphorical arrangement of the universe, among other things, which form and are formed by cosmological principles.

Accordingly, the extraction of ancient Maya cosmological beliefs from their material remains is still sometimes characterised as a tenuous practice of applying contextual ethnohistorical knowledge onto a wide array of earlier material culture (Smith 2003). However, the cultural connection between the Postclassic and earlier Maya does warrant an understanding that the similarities between their material culture represent not so much disparate traditions as they do points along a continuum. Therefore, the use of ethnohistoric (and ethnographic) knowledge, if used critically, does much to illuminate our understanding of the Classic period or, at the least, to generate conceptual models against which the archaeological record can be tested (e.g., Fox and Cook 1996; Vogt 1970). It must be emphasised that the search for meaning is ongoing, and has drawn primarily upon research into iconography, ethnohistory, and analogies drawn from contemporary Maya peoples, such as the Lacandon (Tozzer 1907) and the Tzotzil (Vogt 1969).

Generally, the picture that emerges is a cosmovision which conforms to a recognisable Mesoamerican Ideal (as in Carrasco 1990). According to Ashmore (1991) this incorporates, in various forms, the following traits:

1. A universe which is partitioned into several layers. At its most basic, these layers correspond to the heavens, earth, and an underworld. These layers have differential associations with supernatural entities, power, and social relations.

2. The unifying character of time and space. This refers primarily to the separate layers of the universe and how they are coupled and conflated through the specific movements of celestial bodies and ritual, which is often, but not necessarily, of a temporal nature.

3. A physical manifestation of the axis mundi connects the layers of the universe and provides a conduit for ritual and supernatural travel between the layers.

4. The superimposition of the cardinal directions (including the concept of "centre") over the layers of the universe. The directions have their own supernatural and social associations, which reinforce those of the universal layers.

The above list provides the main (and very basic) structural elements of ancient Maya cosmology. A brief review of these elements will help to elucidate their relation to ancient Maya experiences of landscape, community, and architecture.

The Multilayered Universe
As stated above, Maya cosmology closely resembles that of other Mesoamerican peoples. This is perhaps most evident in the conceptual arrangement of the universe into related but separate existential planes. Berdan (1982:123-124) tells us that the Postclassic Mexica (Aztec) civilisation of Central México believed in the existence of thirteen layers within a heavenly realm, an earthly realm, and nine levels of an underworld, of which the furthest removed from the earth was Mictlan, the "Place of the Dead." Reconstructed murals from the Classic highland Mexican city of Teotihuacan suggest a partitioning of the universe into several layers, including

a Central Mexican view of Paradise (Nicholson 1967), and Schele and Mathews (1998) postulate that Preclassic Olmec iconography from Veracruz, México, alludes to a similar belief.

Ethnohistorically, the Maya of the northern Yucatan Peninsula partitioned the heavens into thirteen layers, each being presided over by a specific deity; collectively, these gods were termed the Oxlahuntiku (Sharer 1994:523). Iconographically, the ancient Maya (during both the Classic and Postclassic periods) represented the earth as the back of a large aquatic creature, usually a caiman, shark, or turtle. This aquatic creature lay atop the thirteen layers of Xibalba, the underworld (Schele and Freidel 1990; Schele and Mathews 1998:37-39). As with the heavens, the various layers of Xibalba were each accorded a supernatural lord, known jointly as the Bolontiku (Sharer 1994:523). Supporting the entire structure of the universe were the four Bacabs, supernatural beings associated with the rain-gods known as the Chaaks (Sharer 1994:520, 532).

Among the ancient Maya, these levels of existence were not an abstract principle removed from social realities. A very real connection was believed to exist between the various layers of the universe, their associated ideological and supernatural aspects, and the operation of society. The heavens were the abode of the ancestors through whom legitimacy was chartered (McAnany 1995, 1998). The underworld was the location where the Mythic Maize Gods (also known as the Hero Twins), Hun-Ahau and Yax-Balam (or, as known to the Postclassic Quiché, Hunahpú and Xbalanqué (Goetz and Morley 1950)), defeated the Lords of Death and brought the world into its present manifestation on 13.0.0.0.0. 4 Ahau 8 Kumk'u, or August 13th, 3114 B.C. (Schele and Mathews 1998:37). The actions of the Maize Gods also set the precedent for ritual bloodletting, the ballgame, and sacrifice as appropriate avenues for power and authority (Schele and Kappelman 2001).

The scholarly acceptance of the ancient Maya as a pacifistic, astronomically oriented society has been soundly debunked for a generation. Nevertheless, the passage of celestial bodies and accordant rituals were intimately tied to ancient Maya understandings of cosmology. One aspect of this was the metaphorical role played by celestial features in representing religious images, such as the portrayal of Orion's Belt as the carapace of the "Cosmic Turtle" through which the Maize Gods escaped Xibalba (Schele and Mathews 1998:37). Extant Postclassic codices deal exclusively with astronomy-based ritual and prophecy (Sharer 1994), and certain architectural types have been demonstrated to align with celestial entities at specific times of the year (Aveni and Hartung 1986). The most obvious examples of such architecture are the "E-Group" complexes present mainly in the Petén region, at such cities as Calakmul, Tikal, and Uaxactun (Folan et al. 1995a; Laporte 1993; Sharer 1994:182). These complexes, which generally consist of three buildings on an elongated platform facing a fourth, eastward oriented structure across a plaza, are believed associated with rituals involving the observation of equinoxes and solstices (although not all such E Groups have demonstrated the proper physical alignment to function as effective observatories, which suggests some other social processes must account for their presence (Aimers 1993)).

The sun, moon, and the planet Venus were particularly important celestial entities in the cosmological structure of the ancient Maya. Representations of Venus are epigraphically connected to the spread of conquest warfare amongst the Classic Maya, especially the "Star-War Complex" of the Petén (involving Tikal, Naranjo, Calakmul, Dos Pilas, and Caracol, among other cities), thought by some to have been imported wholesale from Central México at the close of the Early Classic. Likewise, the movement of the moon has been shown to be the basis for what Morley (1916) termed the "Supplementary Series" calendrical inscriptions.

Just as the passage of time was partially recorded via the movements of the heavens, the completion, or ending, of certain units of time was ritually significant. Katun (7 200 days) and baktun (144 000 days) completions were physically commemorated in historical inscriptions and ritual. An extreme version of this behaviour is the creation of the Twin Pyramid Complexes at Tikal, which were constructed so as to coincide with katun endings (Jones 1969).

The Axis Mundi
Another manner in which the layers of the ancient Maya universe were connected to one another was through various axis mundi. Briefly, an axis mundi is that which spans the various cosmological levels and, as such, is a world-centering ideal (Eliade 1959). For the ancient Maya, the axis mundi was the meeting of the horizontal and vertical dimensions of space: vertical, as already stated, in that it conjoins the layers of the universe, and horizontal in that it is usually the intersection of the cardinal points (discussed below). As illustrated by Freidel et al. (1993), however, the royal ahau could assume a personified form of the axis mundi during ritual occasions (see also Freidel and Suhler 1999). There also existed permanent and more socially accessible means of bridging the divides between the cosmological planes. Caves and mountains, both of which held religious and mythical significance for the ancient (and modern) Maya, formed one such means (Brady and Ashmore 1999; Brady 1997; Carrasco 1990; Schele and Freidel 1990; Schele and Mathews 1998; Vogt 1969).

The mountain, or witz, by virtue of such significance and innate physical properties, formed a metaphorical conduit between the earthly and otherworldly realms. Schele and Mathews (1998), and Schele and Kappelman (2001) have written of the mythic underpinnings of mountain imagery in Maya, and, indeed, broader Mesoamerican culture.

Directionality
Much like the different levels of the universe, the separate cardinal directions (including the concept of

"centre") were imbued with meaning (Coggins 1980). This can be seen symbolically in the imagery of the quincunx, or cross motif, with its five-fold pattern revealing a directional layout (McAnany 1995: 84-86). The paramount (or favoured) position was the east, presumably due to the symbolic importance of the rising sun (Schele and Freidel 1990). Archaeological research by Leventhal (1983) indicates that within the Copán Valley, and at the cities of Tikal and Seibal, this position is predominantly the area (in relation to residential compounds) where family shrines are located (see Becker (1999) for similar conclusions at Tikal, as well as Schwake (2002, 2003) for Minanha'). Ethnohistoric research focusing on the colonial-era Yucatec Maya reveals that it was this direction, rather than north, which it was placed at the top of maps (Marcus 1993).

Conversely, west was an inauspicious direction, associated as it is with the setting of the sun. Harrison (2001) notes that at Tikal there exist no known examples of royal throne structures which would have enjoyed a west-focused view.

Coggins (1980), drawing on the work of Brotherson and Ades (1975; Brotherson 1976), notes that the north-south directional opposition was considered somewhat differently among the ancient Maya than east-west dualism. She argues that although the ancient Maya identified both the north and south directions, they were only conceived of in relation to the east-west opposition (Coggins 1980). That is, rather than being a distinct division of space, north was "in between" east and west, while south was "in between" west and east. This is ultimately based, Coggins (1980) suggests, on the daily circuit of the sun, which is a fairly common cross-cultural means of dividing space (and time) (Pearson and Richards 1994).

For the ancient Maya then, north was conflated with "up", the sun at its zenith, while south is considered "down", corresponding to the night (when the sun is at its furthest from "up") (see Bricker 1981). The cardinal association of such points "in between" east and west results from imposing a three dimensional concept onto a two-dimensional medium, such as a map, or the surface representation of a city layout. By virtue of their associations with verticality, north and south are intimately associated with the first principle covered in this discussion of Maya cosmology, that of the multilayered universe (Ashmore 1991, 1992).

Maya Cities and Cosmology
How then, did these religious and cosmological models influence and shape ancient activities related to architecture and the construction of civic landscapes? According to Ashmore (1991, 1992, 2002) and Coggins (1980), at least, these areas converged in the overall structure of ancient Maya cities. Specifically, they argue that ancient Maya cities exhibit design principles drawn from cosmological concepts, most notably those related to a multilayered universe and directionality. In conjunction with the claims made through architectural iconography (Fash 1992; Freidel and Schele 1988a; Looper 1995; Schele and Mathews 1998; Miller 1998), and archaeological work (Freidel and Suhler 1999; Brady 1997; Driver and Wanyerka 2002) it becomes clear that the ancient Maya expressed their cosmological and religious views, in part, through the ideal of an ordered city.

Many scholars have argued that ancient Mesoamerican cities were arranged so as to be microcosms of their particular view of the universe (Ashmore 1989, 1991, 1992; Carrasco 1990; Cowgill 2000; Kowalski 1999; Orr 2001; Sugiyama 1993). The ancient Maya cities incorporated, to varying degrees, the above-mentioned cosmological ideals, most evident (archaeologically) in the deference to concepts of directionality and the multilayered universe. This is particularly evident in the twin pyramid complexes of Tikal (Jones 1969).

As already stated, the twin pyramid complexes were tied to katun completion ceremonies. Archaeologically, they present a single construction event which was monumental not only in the organisation of labour and materials, but also in terms of the social and political power necessary to implement such programs. The presence of such organisational capabilities has been used to support of the notion of an ancient Maya middle class, one of the hallmarks of a multi-tiered, centralised model of ancient Maya society (Chase et al. 1990; Haviland 1970).

Also monumental and no less complex, argues Ashmore (1989, 1991), is the reliance on, and utilisation of, ancient Maya cosmological principles in the design of the twin pyramid complexes. The eastern and western limits of the complexes are (usually) marked by the placement of four-sided pyramids with radial stairways. These pyramids are separated by a plaza, and terminate in platforms lacking any kind of superstructure (such as temples). To the northern extreme of the plaza is a roofless structure within which resides a stela or stelae commemorating an ancestral figure, ruling monarch, or various combination of the two. Located opposite this building, on the southern terminus of the plaza, is a single-roomed range structure with nine doorways.

Ashmore (1989, 1991) argues that the opposition of the pyramids is a physical reference to the primacy of the east-west directional route. Following Coggins (1980), she feels that the northern structure in the area symbolically denotes "up". As the structure itself is typically roofless, and as upward verticality is associated with the heavenly realms, there is no physical impediment separating the figures alluded to in the stelae from placement in such realms and, hence, they are conflated with the deified ancestors who reside therein. Likewise, the southern structure, with its nine doorways, symbolises the Classic period equivalent of the Nine Gods of the Underworld, the Bolontiku.

A similar example of the incorporation of cosmological and religious ideals within ancient Maya civic plans can be found in the correlation between royal architecture and underground caves at Dos Pilas, Guatemala (Brady 1997). Research carried out under the auspices of Vanderbilt University's Petexbatun Regional Cave Survey revealed there exists a direct, physical connection between a system of underground, water filled-caverns and the placement of the royal acropolis (Brady 1997:605). The acropolis itself (El Duende) was built on top of a hill which lies overtop of the largest concentration of cavern water. Rather than accessing this water directly from the acropolis itself, the ancient Maya made use of alternate cavern entrances, or springs. The act of placing the seat of political authority on this particular spot, then, more likely had to do with incorporating the cosmological symbolism of the mountain and cave, thus lending ritual and religious credence to the ruling faction, and reframing the social order as something cosmologically ordained.

SUMMARY

The utility of testing the Ashmore and Sabloff (2002) proposition lies in providing an opportunity to examine the nature of ancient Maya political structure through architectural means, and hence help elucidate the more likely socio-political model for the ancient Maya, while at the same time accounting for the widespread occurrence of cosmological symbolism as it pertains to political power (as the two are closely related). In order to address such issues, we must review current theoretical concerns pertaining to landscape and political power, and what it means to experience landscape and political power. These issues are the focus of the following chapter.

CHAPTER THREE

THEORETICAL CONSIDERATIONS

The premise of this thesis, as with Ashmore and Sabloff's (2002) claim, rests upon the notion that architecture, as landscape, performs a communicative role. This stance, although relatively new to Mayanists, has a long history in other areas of archaeology, anthropology and, indeed, the social sciences in general (Thomas 2001:165). In order to address this issue, it is necessary to set the context of Ashmore and Sabloff's study within the broader spectrum of landscape archaeology – including the built environment and city planning, as well as semiology and the ties between material culture and metaphor.

LANDSCAPE

Until the integration of "landscape studies" (primarily that of human geography) with anthropology and archaeology in the 1980s (Cosgrove 1988), the study of the built and natural environments was largely conducted through the paradigm of generalised positivism which held sway over mid 20th century social sciences. In large measure, this approach to landscape owed its immediate origins to the neo-evolutionism of White (1959) and the cultural ecology of Steward (1950).

White claimed that culture (C) was the end result of a thermodynamic model which involved the extraction of potential energy (E) from the environment through the application of various technologies (T); hence his formula $E \times T = C$ (Erickson and Murphy 1998: 117). In this sense, land is a collection of resources, through which the manner of a culture's development is determined. This tied into his "layered" model of culture in which the foundation, and prime analytical importance, of any society was its economy and technology, which were indivisible from the land itself, as it is the source of both energy and material:

> "Other factors remaining constant, culture evolves as the amount of energy per capita per year is increased, or as the efficiency of the means of putting the energy to work is increased" (White 1959: 368-369).

For his role in defining the positivist approach to landscape, Steward is credited with bringing forth the notion of "cultural ecology", which focused on the utilisation of culture as an adaptive mechanism to cope with geographical and environmental conditions (Erickson and Murphy 1998: 118). What is implied in Steward's concept of culture's reliance on ecology, is that similar environments will result in similar cultural developments, and this can be used in the categorisation of societal types and the generation of cultural universals (Gosden 1999:89). Although Steward's theories are different from the "neo-evolutionism" of White (Erickson and Murphy 1998:118-9), and should not be read as a form of absolute determinism (Steward 1955:229 n.3), both of these approaches advocated the belief that land, whether as a local or general principle, was something to be objectified, quantified, and which owed its cultural importance to its role as the origin of subsistence and economics (Steward 1950). As Knapp and Ashmore (1999:1) would later remark, the views of Steward and White enshrined in modernist dogma that landscape was "the backdrop" against which culture occurred.

Neo-evolutionism and cultural ecology influenced a generation of archaeologists, most notably Binford (Trigger 1989:295). It was Binford, in his role as the voice of the New Archaeology in the 1960s and 1970s, who advocated moving away from "pre-scientific" or "cultural historical" interpretations in favour of those which were based on a scientific understanding of cultural processes – a processual archaeology.

Binford's role in the development of archaeological thought is monumental and complex, and the present discussion only touches on some of the aspects with which he has been involved. His embrace of White and Steward's ideas is seen in his intellectual stance that culture is "humanity's extrasomatic means of adaptation" (Trigger 1989:298). While he admitted that "humans may reposition their adaptive strategies within a landscape" (Binford 1982), landscape itself was seen as a component of a larger system which affected the material circumstances of society. The systems-theory approach advocated under the umbrella of processual archaeology had intellectual precedence in the functionalist works of Childe (1939), and enjoyed elaboration in the writings of Clarke (1968: 43):

> Cultural systems are integral whole units. Material culture, economic structure, religious dogma, and social organisation are in this hypothesis merely subsystems arbitrarily extracted from their coupled context…the sociocultural system is a unit system in which all the cultural information is a stabilized but constantly changing network of intercommunicating attributes forming a complex whole – a dynamic system.

Clarke goes on to emphasise the Binford approach when he claims that "cultural ecology forms a useful frame for the study of whole sociocultural systems" and "cross-cultural regularity in material culture subsystems may be seen as but one vector expressing the equilibrium between [ecology and culture]" (Clarke 1968:124).

Alternative approaches to studying landscape did exist alongside the general positivist framework, but did not have the same impact on archaeological practice, and remained largely the preoccupation of ethnologists. Lévi-Strauss, for example, implicitly addressed landscape as a symbolic construct whose societal meaning could be derived through a structural analysis of mythology. Thus, his study, "The Story of Asdiwal", ties Tsimshian mythic expressions of landscape to a general cosmological scheme (Lévi-Strauss 1967). While this approach has more in common with the work of later archaeologists studying the cultural meanings of landscape (Dunning et al. 1999; McAnany 1998), Lévi-Strauss held that the cosmological and religious affiliations expressed through structural analyses were reflective of a pan-human, generalised cognitive setup. Thus, his approach to studying landscape was as normative as positivistism was determinative (Giddens 1984:1-2).

With the inevitable shift away from positivist understandings of culture in the late 1970s and early 1980s, and with the influx of ideas from other academic disciplines, came a re-evaluation of the role of landscape in archaeology (Ashmore and Knapp 1999; Cosgrove 1988; Ucko and Layton 1999). Rather than a conglomeration of resources, landscapes came to be studied in much the same manner as had monuments and architecture, with concern for the symbolic and ideological role that they played in the past (Cooney 1999).

Partly as a result of this concern, and partly due to the perceived need to rectify what has been termed a "colonial" approach to anthropological and archaeological research (Trigger 1999), some anthropologists have tried to incorporate indigenous "voices" into their analytical processes, creating alternate frameworks with which to develop interpretations (Hodder 1992:167; Wagner 2000). One effect of this has been a growing awareness amongst social scientists of the plurality of meaning that a landscape may hold for specific (non-academic) audiences, as well as the contestation of such meaning (Sable 2001).

It has been argued, moreover, that in order to fully appreciate the plurality of meaning, it is necessary to attempt to consider landscape from within the conceptual frameworks utilised by the particular peoples themselves. For example, citing the Maori conception of place and time, Smith (1999:50) states that "there are positions within time and space in which people and events are located, but these cannot necessarily be described as distinct categories of thought". Likewise, the ethnographic studies by Vogt (1969) and linguistic research by Watanabe (1983), among the Tzotzil and Mam Maya respectively, offered new information for archaeologists to utilise emic constructs of cosmology and landscape in the pursuit of the ancient Maya past.

The anthropological alternatives to modernist interpretations of landscape are by no means in agreement amongst themselves as to methodology and process, nor do they share a theoretical unity. Variously, these approaches are grouped under the rubric of "postmodernism" or, in archaeology, "post processualism" (Johnson 1999: 98). Their diagnostic features which separate them from the orientations of modernist dialogue are a willingness to engage in alternative types of investigation and the incorporation of previously unconsidered data in constructing their interpretation of landscape (Hodder 1992: 167-174).

The various post-processual (or interpretive) archaeologies which have had the most pronounced impact on landscape studies include the symbolic and textual (Barrett 1999), and phenomenological (Tilley 1994; Thomas 1990). As such, much of the subject matter of landscape archaeology and anthropology deals with issues of meaning:

"the most prominent notions of landscape emphasise its socio-symbolic dimensions: landscape is an entity that exists by virtue of its being perceived, experienced, and contextualised by people" (Knapp and Ashmore 1999:1).

This is precisely the stance taken by archaeologists who endeavour to understand the "built environment" as a meaningful construction of landscape.

THE BUILT ENVIRONMENT

The term 'built environment', as utilised in archaeological discourse, can be defined as the artifactually reordered environment (Presoizi 1979). The subject matter of this definition is quite broad, including monuments, individual buildings, and other single structures, as well as the human–built open areas which separate them. Taken together, these elements constitute large scale built environments such as towns and cities, the settings for an urban landscape.

One route for studying the built environment which hinges upon the integration of such disparate architectural units, and which is perhaps more familiar to Mayanists than others, is the settlement archaeology which characterised much of the archaeological inquiry of the area over the past forty-five years, especially that designed specifically to locate otherwise unknown archaeological sites (Chase 2004, Puleston 1983). Owing its auspicious beginnings to the work of Willey (himself a student of Steward), who pioneered the practice of settlement archaeology in the Viru Valley of Perú before applying it to the ancient Maya settlements of the Belize River Valley (Willey 1953, 1956; Willey et al. 1965), settlement archaeology remained (and in most cases, remains [for example Ford 1991]) largely concerned with the relation of humans to their ecological surroundings and, as such, adhered to the processual generation of broad cultural generalisations. As Chase (2004:321) describes it,

the primary goal of settlement archaeology has been to understand how humans distributed themselves over a given landscape. For the most part, settlement archaeology has incorporated an ecological approach that examines the relationship of environment or natural landscape factors to the physical location and density of human settlement. This, in turn, has led to the development of predictive models for the location of settlements.

This being said, the past thirty years have seen a gradual increase in the number of archaeologists who employ settlement data to answer other, non-ecological questions, mainly those relating to social organisation, and the spatial organisation of political units. Marcus, for example, incorporated settlement pattern data into her analysis of the political sphere of Calakmul, México, and concluded that outlying secondary centres, many of which were connected by sacbeob (ancient Maya roadways) to Calakmul and their own tertiary centres, tended to congregate at an average distance of 30 km from a primary centre (Calakmul), according with a central place model of settlement. The resultant picture of the Calakmul polity then, esembled a "hexagonal lattice" network of secondary and tertiary centres emanating outward from the primary (politically dominant) centre (Marcus 1976). This model could then be applied to other cities in the Maya area to resolve questions of polity size.

Another approach dealing with the same issues had been undertaken by Hammond (1974) in his application of Central Place Theory to Maya cities, which required Theisson polygons to be drawn around ancient Maya centres, the boundaries between such centres placed at the midpoint between neighbouring cities.

In a similar vein, Adams and Jones (1981) constructed a settlement hierarchy for the Maya lowlands based on a volumetric assessment of architectural mass, with the assumption that those cities with greater architectural concentration would have, of necessity, required more political clout in order to erect them. Hence, in this framework, city size equalled labour investment, which in turn relied upon a broad authoritarian base, a scheme quiet similar to White's (1959) formula, $ExT=C$.

As in most settlement archaeology, however (including the work of Ashmore and Sabloff [2002]), these examples were biased in that their examples were almost exclusively based upon the formal analysis of building form, without the benefit of extensive excavation. As a result, there is very little chronological control, and the models themselves remain synchronic, limiting their theoretical applicability to the final occupational phases of the involved cities, primarily the Late Classic (c. 600-900 A.D.) and Terminal Classic periods (c. 900-1000 A.D.) (see Leventhal and Ashmore [2004] for an example of this problem for the upper Belize Valley).

Aside from settlement archaeology, the study of urban landscape has been the subject matter of geographers, sociologists and, from a different approach, classicists (Ashmore 2002; Rykwert 1976; Sanders and Webster 1988; Southall 1998). Although the theoretical concerns and traditions of these sets of scholars largely differed from that of the majority of archaeologists, there have been some influential works which directly influenced the manner of recent urban landscape archaeology. Rykwert, for example, in his seminal study of ancient Rome, applied ancient historical records to discern the religious and cosmological principles employed in the founding and habitation of the city. He made the apt statement that our modern preconceptions of what a city is and how it operates hinder a true understanding of past societies, and that for a full appreciation of ancient urban meaning, scholars had to strive to overcome the "conceptual poverty of our city discourse" (Rykwert 1976:23).

He claimed that the concerns raised by scholars acquainted with a modernist perspective on cities had to be redefined in order to be effective:

The rectilineal patterns of the Roman towns…are thought to be the byproduct of a utilitarian surveying technique. This is not how the Romans themselves saw it: the city was organised according to divine laws [Rykwert 1976:25].

Rykwert's work was a historical rather than an archaeological one, but did incorporate archaeological data, if not the then current archaeological theoretical concerns, into the overall analysis. This integration of text and archaeology echoes the points raised above in the post-processual critique of positivist theory. Recently, archaeologists of the built environment whose research pertains to historical study have combined disparate data sets (including ethnographies and oral histories (Marcus 1983, 1993; Trigger 2003)) to formulate a deeper understanding of their subject matter. Rather than the "conjunctive approach" advocated by Taylor (1948) and perfected under the New Archaeologists, however, these studies have not aimed to produce generalising socio-cultural classificatory statements, but rather to get as close as possible to the emic understanding of their subject matter. This is especially pertinent in the consideration of city-planning.

City Planning
City planning can de defined as "the disposition and organisation of structures and open spaces according to a conceptual order following pre-established rules of functions, ritual, or aesthetics" (Lampl 1968:6). Although city planning has been approached from several stances, what concerns us here is the nature of the "conceptual order" as stated above. In particular, a discussion of the culturally-specific factors involved in city planning and the categorisation of city types will help to elucidate this issue.

As Lampl's (1968:6) definition for the topic alludes, the planning and organisation of cities, amongst ancient civilisations at least (Trigger 2003:541-583), was a multifaceted endeavour designed for the intersection of several areas of societal experience.

The functional dimension of city planning has been addressed extensively by Southall (1998) and Fox (1977). One result of such analysis has led scholars of Mesoamerican city forms to develop a taxonomy of ancient cities (Marcus 1983; Sanders and Webster 1988). As Marcus (1983:196) states, most such classificatory schemes are based on five interrelated traits: size, location, function, position in a hierarchy, and form. Furthermore, most schemes rely upon the application of interpretive constructs derived from sociology, such as the "concentric" model, which is "based on the idea that most cities have a single centre from which growth moves outward, giving rise to concentric zones" (Marcus 1983:199), as well as variations such as the "sector model", and the "multiple nuclei model" (Marcus 1983:200-206). Importantly, Marcus (1983:198) makes the point that "the more one tries to devise a scheme for the ancient city, the more one is forced to ignore important 'exceptions'".

Sanders and Webster (1988), following on the earlier work of Fox (1977), categorise Mesoamerican cities as falling into one of two functional types, "regal-ritual" or "administrative". This dichotomy of urban "types" has been criticised by Chase et al. (1990), both for its simplification of urban complexity in Mesoamerica, and the uncritical application of historically and socially removed interpretive models.

While the designation of ancient cities to functional classes is useful in a heuristic sense, it does little to elucidate the meanings of city plans. Marcus (1983:206) is correct when she claims that "non-Western people may use very different criteria" for defining their cities. Such categorisation imposes modern logistical structures onto the data and, hence, do not aid in revealing emic concerns. Furthermore, as stated by Trigger (2003:541): "[i]n considering the art and architecture of early civilisations, it is impossible to separate functional considerations from symbolic ones". As Ashmore and Sabloff's (2002) thesis deals with an ancient Maya understanding of their built environment, this discussion avoids a strict review of plausibly separate elements of city planning, but rather strives to present (to the best of our current ability) the emic conceptions of what constituted particular built environments.

Pertinent to the present discussion are the ritual and aesthetic "rules" which shaped the societal template of city designs. One such "rule", according to Trigger (2003:120-140), is the institution of kingship, which in ancient states centred not only on the social, political, and economic separation of the ruling stratum from others, but on the notion that the institution itself (if not individual regents) was a sacred one, one closely tied to the maintenance of cosmic order and rituals of communion between the population and their ancestors and/or the gods. Public monuments and architecture then, as ruling and elite stratum projects, were concerned with the expression (and validation) of such concerns as well as the physical promotion of the current social order.

This concern amongst ancient civilisations to make the "political aesthetic" (Smith 1996) has been remarked upon by many scholars who see such activity as a hallmark of the "theatre state" as described in the previous chapter (Demarest 1992; Demarest et al. 2003). However, for our present goals, suffice it to say that in public architecture we find the means of communicating social ideals, even if such ideals themselves reflect a tightly circumscribed segment of a population (Hardy 1968; Marcus 1992).

An example of such activity and concern with proclaiming the sacred charter of rulership through city planning can be found in the medieval Hindu capital of Vijayanagara, in southern India. The city itself is internally demarcated into four distinct areas organised in a quadripartite arrangement, and separated from one another through means of roadways and high walls (Fritz and Michell 1987:114-120). The quadrants themselves have been interpreted as separate public, ceremonial, administrative, and royal residential areas (Fritz 1986; Fritz and Michell 1987:114-120).

Importantly, the point of convergence of these quadrants is the location of the main ritual precinct of the city (focused on a centrally placed temple), in which religious and royal iconography are displayed to the exclusion of all others (Fritz and Michell 1987:114-115). Additionally, the polity's road system, which in antiquity linked Vijayanagara with its dependencies, begins and is centred on this central ritual area (Fritz and Michell 1987:120).

Fritz (1986) ties historical Hindu understandings of kingship into his interpretation of such a city layout. He claims that the ritual centre of Vijayanagara especially, and to a lesser degree the city as a whole, owes its physical properties and social importance to its expression and facilitation of sacred kingship, especially through ritual:

> While kings were not necessarily divine, the institution of kingship was. Royal authority was intimately bound up with divine power. But...it was also tied to material prosperity, moral well-being, and cosmic regulation...The spatial organisation of [the city centre] provides the setting for rites that also directly invoke the roles of the king. Specific features of the site plausibly suggest the enactment of these various roles [Fritz 1986:46-47].

In the spirit of Wheatley (1971, 1979), Fritz (1986:49) concludes that Vijayanagara was, to its inhabitants, a "cosmic city" embodying the role of the king, characterised architecturally by "clear orientation (in alignment with the cosmos), the symbolism of centrality, and the throne of the king."

These studies of Vijayanagara offer interesting parallels with, and a pertinent point of departure into, a discussion of ancient Maya city planning and aesthetics. Like the medieval Hindu capital, ancient Maya cities seem spatially focused around elite and, more pointedly, royal architecture (Christie 2003; Houston 1998; Inomata and Houston 2001a, 2001b). Similar to the medieval Hindu rulers, it was the integrative role that the royalty played between the various elements of their culture's cosmology and religion which defined the nature of such architecture (Demarest et al. 2003). Such architecture was, furthermore, conceived and portrayed through particular aesthetic concerns, to which I now turn.

Mesoamerican Civic Aesthetics. It has been claimed that the general similarities in physical form between ancient Maya cities and those of their Mesoamerican neighbours results from adherence to shared cultural aesthetics, most often presented in the form of an aesthetic trope, based upon a pan-Mesoamerican cosmological tradition (Carrasco 1990; Joyce 2000; Kowalski 1999; Reese-Taylor and Koontz 2001; Schele and Kappelman 2001). An aesthetic trope can be defined as that which has the ability to transmute core cultural patterns through formal metaphoric properties, modalities, and media, thus providing the structuring principles for any socially effective communication....Aesthetic tropes encode the metacommunicative patterns of a people [Reese-Taylor and Koontz 2001:3-4; emphasis in original].

Schele and Kappelman (2001) claim that the primary aesthetic tropes pertaining to civic planning and architecture in Mesoamerica are those which deal with the mythical origins of war and sacrifice as legitimate elite means of maintaining the cosmic order, as well as the location of the primordial act of creation. The setting in which the mythical events took place was the location of Tollan (not to be confused with the seat of the Early Postclassic Toltec polity now recognised as Tula, Hidalgo, Mexico), a location whose physical properties provided the architectural means and symbols of authority, and as such, a cultural template for the promotion and reification of such authority. Specifically, Tollan consisted of a sacred mountain, which the Mexica termed Coatépec and the Classic Maya Kan-Witz ("Snake Mountain" in both cases), at the base of which was a ballcourt, and which projected up from the centre of a lake (Schele and Kappelman 2001:43). Conflated with the notion of Snake Mountain, furthermore, is the locale where the gods emerged from the earth with maize for the first humans, referred to as Yax-Hal-Witz, the "first true mountain", also known as "Sustenance Mountain", and through which communion with the ancestors and gods was facilitated (Schele and Kappelman 2001:38; Schele and Mathews 1998:48).

The sanctified activities of war and sacrifice initiated at Snake Mountain and the provision of maize (the fundamental basis of Mesoamerican civilisations) created a reciprocal obligation between society and the supernatural, a relation referred to by Arthur Joyce (1999) as the "sacred covenant" of Mesoamerican civilisations; this provided the framework for the appropriation of cosmological imagery and authority by the elite. It was this understanding, in its various regional expressions, which provided the underlying message of Mesoamerican architecture and city planning (Kowalski 1999).

Amongst the ancient Maya, Sustenance and Snake Mountains were signified iconographically. These mountains and their associated caves were represented in the form of the "witz monster", a supernatural being often in the form of a snake or a jaguar whose typical open maw is believed to have represented a stylised cave, and hence, a portal to the underworld. The witz monster motif often adorned the facades of temple-pyramids, denoting their cosmological associations, such as on Structure H-X-Sub 3 at Uaxactun, Structure B20-2nd at Caracol, and Tikal's Structure 5D-33-2nd (D. Chase and A. Chase 1998; Schele and Kappelman 2001:39-40; Schele and Mathews 1998:43). Non-Maya examples of this imagery can be found in the Temple of the Feathered Serpent at Teotihuacán (Cowgill 2000; Schele and Kappelman 2001:38; Sugiyama 1993) and the Templo Mayor at Tenochtitlán (Carrasco 1990; Carrasco et al. 1988).

Another means of stressing supernatural and, in this case, specifically underworld locales, was through the creation of pib nah, or "underground houses/structures" (Schele and Mathews 1998:40). This was denoted at Tikal and Palenque, for example, through the application of Itzam-Ye carvings to the entablatures of such buildings. These avian creatures are thought to be connected to Itzamna (the "smoking god"), a mythical personage tied to the Maya Hero Twins (Schele and Mathews 1998:36-40, 84).

In addition to iconography, (which was in many cases restricted to the tightly circumscribed inner sanctums of temples and royal acropolises (McAnany and Plank 2001)), the ancient Maya elite, at Tikal in particular, augmented the validity of these aesthetic tropes, and hence of their social positions, through an application of cosmological understandings of directionality. Similar to that employed in the Twin Pyramid Group Complex discussed in chapter two, the Tikal elite manipulated meaningful space so as to symbolically align themselves and their ancestors with the heavenly realm, on the one hand, and with the creative forces of Xibalba on the other (Ashmore 1989, 1991). In her extrapolation of the Twin Pyramid Group Complex onto the whole of the Tikal epicentre, Ashmore (1989, 1991) claims that the northern, roofless structure which marked the northern limit of the complex was replaced by the North Acropolis, the area in which the majority of Tikal's ancestral elite were buried, symbolically placing the royal dead in the upperworld. The east and west temples are found in Temple 1 and Temple 2 respectively, separated by the Great Plaza, and the southerly-placed range structure with nine doors corresponds to the palace compounds of the Central Acropolis, specifically Structure 5D-120, which has nine such openings. The pairing of the elite residence with the Bolontiku implied that access to the underworld (and hence, authority) was derived through the sovereign, and reinforced the notion of a divinely sanctioned kingship (Freidel and Suhler 1999).

The conflation of ancestors, especially those belonging to the elite stratum of ancient Maya society, with the Oxlahuntiku, the gods of heaven, illustrates the ancient Maya practice of apotheosising their forbearers (Fash 1998). It has been extensively and persuasively argued by McAnany (1995, 1998) that such behaviour forms the core of ancient Maya religion and political authority. She states:

> Communicating with deceased progenitors was not a religious experience divorced from political and economic realities...rather, it was a practice grounded in pragmatism that drew power from the past, legitimised the current state of affairs (including all the inequalities in rights and privileges), and chartered a course for the future. Ancestors resided at that critical nexus between past and future, and their presence both materially and symbolically lent weight to the claims of their mere mortal descendants [McAnany 1995:1].

Ancestral veneration, forming as it does the basis of ancient Maya religious practice, was the core of elite ritual (Miller 1998; Schele and Freidel 1990). As the ancestors of the elite were symbolically divine, the ultimate origin for the political apparatus of ancient Maya kingship was found in the aesthetic tropes discussed above (Freidel and Schele 1988b:548). Compounded with the ancient Maya appreciation of time as a cyclical phenomenon which unified the various levels of the universe (as stated above), the reigning royal elite during ritual activity may have been viewed as a collective incarnation of the Maize Gods/Hero Twins, the mythical actors tied to Sustenance Mountain (Freidel and Schele 1988b).

Ritual activity was facilitated through elite architecture. Freidel and Suhler (1999) have shown that Late Preclassic buildings at Yaxuna, as well as Classic period palace compounds at Copán and Palenque, were designed so as to allow ritual activities which replicated the solar circuit – the cosmic metaphor for the trials of the Maize Gods/Hero Twins, who were iconographically related to both the sun and Venus. This is in accordance with work by Freidel and Schele (1988a, 1988b), which claims that the central cosmic issues of death and rebirth were enacted through elite-focused ritual in conjunction with specific architecture at Cerros, which involved a degree of public spectacle. As Miller (1998) argues for the public ritual of Copán, it was within such dramatic frameworks, rather than through the esoteric iconography and concepts of directionality, that the majority of the ancient Maya would have experienced and understood civic architecture.

The utilisation of aesthetic tropes as a means of communication and social validation, and the notion of praxis, necessitates a discussion of the how the meanings of architecture are convincingly communicated.

Architecture as Communication
As stated at the outset of this chapter, the integration and subsequent analysis of Ashmore and Sabloff's (2002) claim with ancient Maya political history requires that the built environment engender some means of communication. In effect, they are arguing for a semiotic understanding of ancient Maya city layouts. Georges Mounin (1980:491) presents a discussion of the fundamentals of semiology which is worth citing at some length:

> The word 'Semiology' currently has at least two meanings. On the one hand, it means the scientific study of all human communication systems except natural language…On the other hand, it signifies the study of all human phenomena to the extent that they have anthropological meaning…The opposition between these two semiologies arises not from a quibble over words, but from the very different nature of the phenomena they study. The semiology of communication studies codes instituted to produce messages consciously sent and received as such…One cannot send a signal, a sign, or a symbol without doing so on purpose; one cannot receive one without knowing that it was purposefully sent to be received as a message. In contrast, the semiology of meaning works with very different phenomena – indices or symptoms. They are phenomena that are not produced purposefully as messages…One interprets (often with difficulty) indicators, lost in a mass of phenomena, while one decodes signs learned purposefully and socially as such. The decoding of messages presupposes the learning of a code; the difficult discrimination and interpretation of indications in a given domain constitute the science of that domain [emphasis in original].

Clearly, what we are dealing with in the city planning of ancient Maya cities is the former kind of semiology, that which is based on interpretation of meaningful indices (as linked to the above-mentioned aesthetic tropes). However, Mounin simplifies the issue by the rigid dichotomisation of semiologies, especially in relation to Bordieu's (1977) notion of praxis, in which the "lived experience" of phenomena generates contextualised social meaning and understandings without, necessarily, premeditated intention, or a formalised plan. Like structuration theory, praxis allocates the determination of meaning to the dialectical relationship between knowledgeable agents and the existing social structures in which they operate (Giddens 1984:3, 376).

This is precisely the point made by Carl (2000) in his discussion of "city-image". He argues that "the order of praxis is the basis of all forms of ordering, even the most radically conceptual" (Carl 2000:334). Contrary to Carl (2000:328) and Mounin (1980:493), however, I do not feel that acknowledging the importance of praxis negates the applicability of aesthetics for understanding meaning couched in expressive (i.e., architectural) terms. Rather, I espouse the view shared by Giddens (1984) that a "lived" understanding and purposeful meaning (as vested in architecture) exist simultaneously, the former filtered through the later and subject to context (Cowgill 2000; Hodder 1987b).

Such meaning, for our purposes, is most readily addressed through analysing the syntagmatic and paradigmatic dimensions of ancient Maya city components. Giddens (1984:17) states:
> In analysing social relations we have to acknowledge both a syntagmatic dimension, the patterning of social relations in time-space involving the reproduction of situated practices, and a paradigmatic dimension, involving a virtual order of 'modes of structuring' recursively implicated in such a reproduction.

Ian Hodder (1987b:3) reiterates the importance of defining and evaluating syntagmatic and paradigmatic contexts:
> In the study of the symbolic order, structure refers to an organised set of similarities and differences. For example, syntagmatic relations refer to sequential groupings. Thus the human body might be divided into a number of parts (head, torso with arms, legs, etc.)…and a settlement into various categories of

space...In contrast, a paradigmatic set is two or more signs, each or all of which can occur in the same position. Membership in the set helps to determine the identity of the sign through difference and contrast [emphasis in original].

The demarcation of entire city units into recognisable syntagmatic and paradigmatic units is a necessary undertaking if the meanings of city plans are to be revealed, and statements made on their adherence to reconstructed ancient Maya political history. Such an approach has been implicitly undertaken by Ashmore (1989, 1991) and, from a different approach, Houk (2003), for the analysis of ancient Maya city layouts. As will be addressed in the next chapter, this study will employ the methodology of both Ashmore (1991) and Houk (2003) to determine and examine the syntagmatic and paradigmatic elements of ancient Maya city plans. An important point of these ideas and, also, of their relation to praxis is made by Jeff Kowalski (1999:12):

Although the plans of sites and designs of adornments of individual buildings may refer to phenomena outside of themselves (i.e., myths, legends, historical events, astronomical events, etc.), they are also inherently "meaningful" insofar as they create a series of visual self-images for members of that society.

SUMMARY

The archaeological study of city planning is, like most endeavours within the discipline, one with a lengthy and complicated history. For the majority of its history, the analysis of ancient cities was subject to the same functional considerations and ecological concerns as was much of the subject matter of positivist anthropology. The application of alternate ways of examining the past, due in large measure to the shift to postmodern anthropological thinking in the latter decades of the last century, has brought about a concern with meaning in ancient city planning, and a focus upon the intentional communicative properties of the built environment. Historical, ethnographic and, importantly for the ancient Maya, iconographic sources have done much to bring these meanings to the fore of archaeological inquiry.

In the study of the ancient Maya built environments, such issues have been effectively explored through the application of cosmologically-derived aesthetic tropes, and the analysis of the syntagmatic and paradigmatic dimensions of the built environment (with an awareness of the social experiences of such settings) through which such aesthetic tropes conveyed meanings. It is to these elements of ancient Maya city plans that I now turn in the presentation and analysis of the data.

are concerned. The architectural data refer to what can be deemed the long-term continuities, while the political situation of the Late Classic can best be described as an amalgam of medium-term and episodic events (Iannone 2002:74; Hodder 1999:130). By this I mean that architecture, as a pervasive characteristic of the urban landscape, might well outlast the shifting political structures with which we are attempting to correlate. Therefore, the issue of the accuracy of the entire exercise is called into question. However, without the subsequent aid of extensive excavation, we must rely on the assumption that the final architectural phases of the Classic period lowland cities (mostly pertaining to the Late and Terminal Classic periods [Ashmore and Sabloff 2002, 2003]) are synonymous with Late Classic political writings.

POLITICAL STRUCTURE

The reconstruction of ancient Maya political behaviour, although greatly aided through the increasing comprehension of their textual records, still presents a formidable and contested challenge to academics. Several generations and many paradigms removed from Morley (1935) and Thompson's (1954) claim that the Maya were a pacifist theocracy, scholars are still in disagreement over the exact nature of their political structure. As discussed in Chapter 2, Martin and Grube (1994, 2000) have recently advocated a "macro-regional" level of socio-political integration for the Late Classic Maya. Through the analysis of epigraphic texts, these scholars have delineated groups of ancient Maya polities which they believe correspond to two such "superstate" organisations, headed by Tikal and Calakmul, respectively. According to Grube (2000) these superstates were extensive hegemonic city-states, in that while the associated polities retained their internal autonomy, their "foreign policy" and relations with other city states were regulated, if not outright controlled, by the dominant city.

As with any aspect relating to ancient Maya epigraphy, the politically relevant information is neither forthcoming nor free from divided interpretations as to its meaning. Partly as a result of this are the divergent stances of those scholars who see the ancient Maya as comprising city-state or regional state level polities, respectively, as discussed previously.

Also, the framework within which we view ancient Maya politics, like the rest of ancient Maya culture, suffers from a lack of consistency. For example, this thesis considers the political atmosphere of the Late Classic period as the pretext for the exercise of exploring site plan similarities. However, the political situation of the Late Classic period is founded entirely on the events of the Early Classic period (300 A.D. – 600 A.D.). This can be most readily seen in the expansion of the Calakmul royal court, which began its program of military expansion, patronage, and long-distance ritual networks as early as 529 A.D. (Martin and Grube 2000:103-104). The ties which Calakmul forged with other cities during this time and its military humbling of Tikal in 562 A.D (together with Caracol), herald the emergence of a new political order in the Maya area, one that is present at least fifty years before the start of the Late Classic period (Marin and Grube 2000:104; Pincemin et al. 1998). As such, it must be understood that the political situation explained here is a disjointed picture of broader socio-cultural processes which have their ultimate origins in the changes of the Late Preclassic (400 B.C – 300 A.D.), and which extend well into the Terminal Classic period (900 A.D. –1000) and beyond (Freidel and Schele 1988b; Henderson 1997:201-227). This is a necessary point to emphasise, lest we find ourselves in the position of methodologically and intellectually segmenting the political activity of the Late Classic as something removed, and disconnected, from the remainder of ancient Maya politics.

The chronological point at which the power-blocs of Calakmul and Tikal were effectively defunct is 849 A.D. This is owing to the fact that the ahauob of both Tikal and Calakmul appear to have participated in the same ritual event marking the 10.1.0.0.0 katun ending at Seibal at this time, underscoring the demise in their political contestation and, hence, their respective "superstates" (Martin and Grube 2000:53, 115). Likewise, it is after this point that we begin to see the erection of stelae at Jimbal, Ixlu, and, once more, at Uaxactun (Martin and Grube 2000:53). Jimbal and Ixlu's proximity to Tikal and their respective silence during much of the Late Classic attest that they were beholden to the larger city until this time. Martin and Grube (2000) list 78 separate ancient Maya cities which may have participated in the superstates of Calakmul and Tikal (excluding the capitals themselves). Of these 78, 28 are only known through hieroglyphic texts; as they have no archaeological representation, they are necessarily excluded from the study. Of the remaining 50 cities, only 47 display primary or secondary degrees of affiliation with Tikal or Calakmul, with 3 cities claiming rotating allegiance (see Tables 1 and 2). By a primary degree of affiliation, I mean those cities that are linked epigraphically to either Calakmul or Tikal, while secondary affiliates are those cities epigraphically linked to the primary affiliates. Of these 47, published maps were available for 34 (see Table 3).

SITE MAPS / ANALYSIS

A perusal of the syntagmatic and paradigmatic elements found throughout the analysis of the selected cities reveals an eclectic mix of Tikal and Calakmul traits. Most cities contain elements from both of these templates, and only in a few instances can it be said that a whole "pre-packaged" civic layout was imposed or influenced by the political clout of Tikal or Calakmul. A detailed review of the paradigmatic and syntagmatic elements will clarify the issue, and place Ashmore and Sabloff's (2002) proposition in context.

Paradigmatic Elements
Tikal Paradigmatic Element 1 / Calakmul Paradigmatic

Element 2 (North-South city orientation). As Ashmore (1991) stated, and as Schele and Freidel (1990) illustrate via iconographic data, the Classic period Maya cities were characterised by the predominance of a north-south civic alignment. Significantly, this is also the case in the present study. Of the 33 cities analysed, 19 (58%) were arranged in such a fashion. Furthermore, many of the cities which do not conform to a north-south orientation seem victims of their topography. Yaxchilán, for example, which is pressed between the westward-flowing Usamacinta River and slopping uplands, precludes a north-south arrangement. The near ubiquity of this trait suggests that it is a pan-Maya civic feature – of the cities analysed, both Tikal and Calakmul affiliates feature this particular element.

Tikal Paradigmatic Element 2 (Twin Pyramid Groups). As previously stated, the civic plan of Tikal has been argued to have been an elaboration of this particularly Petén architectural arrangement. Of the cities studied, only one site, Yaxha, evinced this feature. Significantly, Yaxha lies within a few day's walk of Tikal, and is considered a secondary affiliate of the larger centre (and, at times, of Naranjo). Therefore, the presence of Twin Pyramid Groups in this city may represent a regional architectural tradition as well as Tikal's political orbit. For these reasons, it is significant that no other city can lay claim to such an arrangement.

Tikal Paradigmatic Element 3 (Formal and Functional Dualism Between the North and South). Assessing the functional nature of architecture solely from city maps is, as previously discussed, a challenging task. The only confident example of functional dualism encountered in the analysed cities is at the secondary Calakmul affiliate Xunantunich which, owing to the relative simplicity of its epicentral layout, evinces such a pattern. The south of the centre is dominated by the (presumably) ceremonial structure El Castillo, while the north features the royal residential area. Significantly, Xunantunich is one of the sites which appears to have been particularly influenced by the layout of Calakmul. Again, this Tikal paradigmatic element may infer a more inclusive practice.

Tikal Paradigmatic Element 4 (The Addition of Architectural Elements on the East and West to Form a Triangle with the North). Two examples exist which feature this building element. Again, the total acceptance of this element is hindered by perspective. The "addition" of eastern and western architectural elements implies a fixed building chronology, something which is beyond the scope of the present study. Nevertheless, both Palenque and Pomona share this feature.

Calakmul Paradigmatic Element 1 (A Southern Pyramidal Structure Facing a Medial Plaza, an E-Group, and a Northern Situated Elite Residential Structure). This element refers to what is basically the entire layout of Calakmul. Three sites in particular participate in such an arrangement: Nadzcaan, Naranjo, and Xunantunich. Significantly, these sites all fall within the Calakmul political sphere of the Late Classic period.

Syntagmatic Elements

Tikal Syntagmatic Element 1 (Transitional ballcourt). As in much of Mesoamerica, ballcourts served religious, ritual, and political ends. The ballcourt itself, as with much Classic Maya public architecture, has its conceptual origins in the pan-mythic and religious conventions of Mesoamerica (Scarborough and Wilcox 1991). Among those cities analysed, 20 shared this feature. As with the orientation of Classic period cities, such numbers are irrespective of political concerns, and cannot be used to support the contention that architectural layout and political power-blocs were connected.

Tikal Syntagmatic Element 2 (Northern-situated, Southern-facing Pyramidal Structure). Of the six cities which include such structures, two are primary Tikal affiliates (Palenque and Uaxactun), two are primary Calakmul affiliates (Caracol and Piedras Negras), while the remaining two are a secondary Tikal affiliate (Nim Li Punit), and an ambiguous polity (Yaxchilán).

Tikal Syntagmatic Element 3 (Opposing Eastern and Western Pyramidal Structures). The opposing eastern and western pyramidal structures compose part of the Twin Pyramid Group complexes at Tikal. In total, seven cities analysed displayed such an arrangement. Five cities were primary affiliates of Tikal, one of Calakmul, while another was a secondary Calakmul affiliate. The spatial distribution of this trait is interesting, for it occurs across the central lowlands, and is found in the eastern and western extremities of the Classic Maya realm. Plausibly, this trait could be the result of a localised manifestation of a shared pan-Maya architectural norm. However, given that it appears with Tikal affiliates as widely spaced as Palenque and Copán, it would seem that if it were such a manifestation, there would be more numerous instances of it in other cities. With that in mind, we must concede the possibility that this particular trait had some political significance to Tikal and her allies.

Tikal Syntagmatic Element 4 (Southern Acropolis with Underworld Imagery). Ashmore (1991:200) describes this element as a southern-situated acropolis in concordance with Underworld iconography or imagery. However, such details are not necessarily accessible through site maps, so all instances of southern-placed acropolises were considered for this particular element. In all, 10 cities shared this design element, 4 of which are either primary or secondary Tikal affiliates.

Tikal Syntagmatic Element 5 (Radial Pyramidal Structures). Like the opposing pyramidal structures, the 4-sided pyramidal structure is an integral component of the Twin Pyramid Group complexes. It does occur in other contexts, however, such as in the western component of E-Groups (for example at Uaxactún and Tikal), and separately (as at Copán). Five cities displayed this particular element: Copán, Uaxactún, La Muñeca, Holmul, and Yaxha. With the exception of Copán, these cities are all located in the central Petén heartland of the Classic Maya, and represent both Tikal and Calakmul affiliates.

Tikal Syntagmatic Element 6 (Northern, Roofless Twin Pyramid Group Structure). This building element corresponded to the ritually important act of conflating the royal ahau with the heavens and, hence, should rate high as an example of political promotion. However, only 1 city, Yaxha, can claim this particular element.

Tikal Syntagmatic Element 7 (Internal Causeways). Although the internal causeway system of Tikal is stressed as a key element of "Petén-style layouts" (Houk 2003) it is, much like the presence of ballcourts, not something defined by geography or, evidently, politics. Of the studied cities, six included sacbeob among their site plans. With the exception of Uaxactún and Yaxha, the remaining cities represent either primary or secondary Calakmul affiliates.

Calakmul Syntagmatic Element 1 (Large Structure 2-like Building). The epicentre of Calakmul is dominated by Structure 2, the large pyramidal complex located in the south of the epicentre, and evidently of Preclassic origins. Regardless of its antiquity (or perhaps because of it), this particular locale became one of the key ritual and political foci of Late Classic Calakmul (Martin 2001; Martin and Grube 2000:107). Ten cities include a large pyramidal structure such as Calakmul's' Structure 2 (with the exception of Caracol's Caana, these buildings do not approach the size of Structure 2). Although Structure 2 is placed in the south of Calakmul's epicentre, both the location of large pyramidal structures, and their absolute presence were noted. Significantly, of the 10 cities which included such a structure, all were either primary or secondary political affiliates of Calakmul.

Calakmul Syntagmatic Element 2 (A North-south Oriented Plaza with a Medial Structure). Eleven cities in this study display this particular trait. Again, there are conflicting data: of the 11 cities there are three secondary Tikal affiliates, two ambiguous polities, while the remainder are Calakmul affiliates (including one primary affiliate).

Calakmul Syntagmatic Element 3 (A Main Plaza Framed by an E-Group). A distinctive trait of Calakmul is that its E-Group, unlike the commonly cited example at Uaxactún, is directly integrated into the core of the epicentre, framing, as it were, the main public plaza of that city. Ten cities share this configuration, either with the main central plaza, or, if such an arrangement is not present, with one of the principle centrally-located plazas. With the exception of Jimbal and Yaxha, all of the included cities are Calakmul affiliates.

Calakmul Syntagmatic Element 4 (A Northern Elite Residence Raised Above the Main Plaza). This architectural element, which is the structural opposite of Tikal syntagmatic element 4, is shared by 11 cities, the most widespread of the Calakmul syntagmatic elements in this study. Of these 11, all are either primary or secondary affiliates of Calakmul, which underscores the political dimension of this arrangement.

SUMMARY

This brief review of the architectural elements present in the cities under study makes apparent the eclectic nature of ancient Maya civic planning. With a few notable exceptions, there seems to be no readily forthcoming correlation between political affiliation and total civic arrangement. What could be approached as an either/or situation (in that the forms indicate exclusive Calakmul or Tikal affiliation) appears more complex, and in need of explanation. The influences, and implications, of such a situation, are addressed in the following chapter.

CHAPTER 5

INTERPRETATION

The correlation between the civic layouts and political structure of the Late Classic appears to illustrate a more complex social process or processes than the ascription of political control. Of the 78 cities analysed, not a single one shared all of Tikal's paradigmatic traits, while only three shared those of Calakmul. This is complicated by the unavoidable fact that Tikal has four paradigmatic elements while Calakmul has only two. Furthermore, Tikal and Calakmul share a north-south civic alignment as one of their paradigmatic elements (Calakmul paradigmatic element two and Tikal paradigmatic element 1).

In terms of syntagmatic elements, the most featured in a single city is six, at Yaxha which, as addressed earlier, may result from a shared regional tradition rather than political control. Of Calakmul's syntagmatic traits, only a secondary city, Xunantunich, ascribes to the entire setup.

Given this information, we cannot assert that there is, in fact, overt political influence in the arrangement of civic areas, as a common feature of ancient Maya society. However, simply negating this idea does not account for the entirety of the data. For example, although no site can lay claim to the entirety of the key diagnostic elements found in the civic layouts of Tikal and Calakmul, almost all of the cities under investigation evince a north-south general alignment, and have medially-placed ballcourts. Furthermore, all of the syntagmatic and paradigmatic elements of the Calakmul and Tikal civic plans are expressed at least once throughout the studied cities.

What then does this mean? Are we witnessing in these civic plans a collection of common Classic Maya architectural traits, or is there a unifying principle to such elements? In order to elucidate these issues, we must follow a hermeneutic approach and, again, review this data in light of its conceptual origins and social context.

Common Elements

If we concede that syntagmatic elements relate to the constituents of a concept (in this case, the ancient Maya city), and that substitutions of such elements results in a new (paradigmatic) understanding of said concept (as in whole-part relations [Hodder 1982, 1999:130-131]), it can be reasoned that, in the case of Tikal and Calakmul, elements common to both their affiliates (such as Tikal syntagmatic elements 1, 2, 4, and 7) were directed to different ends. That is, the paradigmatic architectural elements were employed to construct different civic templates in the two polities. This, in turn, implies that the various syntagmatic elements employed were themselves of general importance. The ultimate origin of these elements can, in fact, be directly traced to the aforementioned cosmological and directional association of such traits.

The ballcourt, itself a means of negotiation between this world and the Otherworld, conflated with its medial placement in the centre of city layouts is a shared feature of Classic Maya ritual and religious concern (Seibert 1999; Scarborough and Wilcox 1991).

Likewise, the northern-situated pyramidal structure can be seen as a Classic Maya representation of what the Postclassic Mexica termed Coatepec (Schele and Kappelman 2001). Even without the iconographic information which may indicate that the analysed cities designated these structures as Kan-witz, the primordial mountain tied to the creation myth of the ancient Maya (Schele and Freidel 1990; Schele and Kappelmen 2001), their northern placement indicates that such structures were deemed an axis mundi, bridging the common world with the heavens (Ashmore 1989; Coggins 1980). The same argument can be applied to the southern-placed acropolises which, by virtue of their placement on the civic landscape, recall the directionally-based associations with Xibalba (Ashmore 1991:200).

It would appear, based on these shared elements, that both Calakmul and Tikal adhered to common culturally-specific aesthetics in choosing the architectural components of their respective cities, as did the other polities.

Politically-Specific Elements

There are, however, a number of traits which appear to correspond with the individual superstate arrangements. Interestingly, more traits are exclusive to Calakmul and her affiliates than to the Tikal superstate. Again, this interpretation is based on the "fitting" of the various bodies of evidence, and does not attempt a quantitative reckoning. Those Calakmul traits to which I refer are Calakmul paradigmatic element 1, and Calakmul syntagmatic elements 1 and 4.

Calakmul Paradigmatic Element 1. As described in the previous chapter, this particular collection of traits replicates the entirety of the Calakmul civic layout. Nadzcaan and Naranjo, both primary affiliates of Calakmul, share this pattern. Nadzcaan, which lies in Campeche, México, is a close neighbour of Calakmul, and it may be possible to see in this relationship the effects of a regional architectural tradition (such as is assumed to account for the occurrence of the Twin Pyramid Group complexes at both Tikal and Yaxha). However, Naranjo is located far to the southeast of the larger city. Naranjo first became a Calakmul dependency during the reign of its ruler Aj Wosal in 546 A.D. (Martin and Grube 2000:71-2), and would remain within Calakmul's political orbit until the decline of Classic Lowland city-state culture.

Likewise, Xunantunich, located in the Belize River valley, which enjoyed a quite brief fluorescence at the beginning of the Terminal Classic period, shares this agglomeration of design elements. Xunatunich is deemed a secondary Calakmul affiliate, representing a client kingdom within the Naranjo political sphere (Leventhal and Ashmore 2004). Interestingly, the majority of the present architecture at this site coincides with the decline of central authority at both Naranjo and Calakmul. Whether the site-plan of this centre results from Naranjo or Calakmul influence, it probably participated in the same structured relations which accounted for the spread of the same design from Calakmul to Nadzcaan and Naranjo.

Calakmul Syntagmatic Element 1. The replication of Structure 2 of Calakmul does seem to have political considerations. As indicated, this structure, like other Maya pyramidal structures, has its conceptual origins in the shared Maya idea of an axis mundi (Ashmore 1991:200). However, of the 10 cities which incorporated this design element into their city plans, nine were either primary affiliates of Calakmul (Dos Pilas, Caracol, Piedras Negras, Quirigua, La Muñeca) or secondary affiliates of Calakmul (Xunantunich, Tamarandito, Ucanal, Arroyo de Piedra). The remaining city, Tonina, is slated as ambigious, but did have relations of some kind with Calakmul (Martin and Grube 2000:184). This is a strong indication that significant political influence accompanied erection of such structures. Significantly, the one exception to the Calakmul monopoly of this element is not a Tikal affiliate, suggesting the exclusive character of this particular trait.

Calakmul Syntagmatic Element 4. The cities which have a raised, northern elite residence are as follows: Naranjo, Caracol, Piedras Negras, Nadzcaan, La Muñeca (all primary Calakmul affiliates), Arroyo de Piedra, Tamarandito, Holmul, Xunantunich, Ucanal, and Cahal Pichik (all secondary Calakmul affiliates). More than any other element this seems to have important political implications, owing to the unanimous correlation of the architectural traits with the political history.

Tikal Syntagmatic Element 3. The only syntagmatic element for Tikal which may indicate a political dimension is the opposing eastern and western pyramidal structures, a component of the Twin Pyramid Group complexes. Although not all of the seven sites which shared this architectural arrangement were Tikal affiliates (Ixkun and Piedras Negras are secondary and primary Calakmul affiliates, respectively; the remainder are Palenque, Copán, Jimbal, Uaxactun, and Yaxha, all primary Tikal affiliates), the presence of this feature in such widely separated clients as Palenque and Copán may indicate that some political importance can be attached to it.

Ancient Ritual Integration: A Possible Explanation
If the elements described above are, in some way, indicative of political structures or ritual activity, what were the means by which they operated? How can we infer their place in the political schemes of Late Classic Calakmul and Tikal? A useful example of how to bridge the archaeological data with its presumed social context is provided by Reese-Taylor's (2002) study of ancient Maya ritual circuits.

Utilising ethnohistoric and archaeological evidence, Reese-Taylor demonstrates that three ritual activities, circumambulation, periphery to centre processions, and base to summit of mountain ceremonies, noted today in the highlands of Guatemala may have precedent throughout the Preclassic and Classic periods in the Maya lowlands. Specifically, she examines the role of ritual circuit-related architecture as a principle factor in ancient Maya civic layouts. Among the historic Maya, circumambulation rituals serve to define the territories of particular communities, serving to unite and demarcate local geographies (Reese-Taylor 2002:145). Like the other aspects of current Maya concepts of space and time, these rituals are steeped in religious and mythic understandings of natural and celestial phenomena (see Watanabe 1983).

Interestingly, the same behaviour is alluded to in Classic period historical texts, although usually concerning the nature of the universe rather than the individual community. At Palenque, for example, texts referring to the creation of the universe describe a circumambulatory pattern (Reese-Taylor 2002:147).

Reese-Taylor (2002:147) argues that the integration of Maya religion and politics which appears in the archaeological record by the Late Preclassic (most notably in the figure of the divine ahau (Freidel 1992; Martin 2001:169)), is evinced in the architectural arrangement of certain building forms at Cerros and Uaxactún, as well as the North Acropolis at Tikal. Essentially, these building forms created public space amenable to circumambulatory rituals which, like their modern counterparts, would have served as integrative devices on both the geographical and community level. Through such rituals, Classic period rulers and nobility could assert their polity's political reach, sanctioned through cosmological understandings.

Another modern ritual which has ancient precedence is the act of integrating periphery and centre through processions (Reese-Taylor 2002:152). These ritual processions are especially important when considering the role of sacbeob in ancient Maya civic designs (Chase and Chase 2001b; Shaw 2001). This accords with the ethnohistorical information on sacbeob rituals from Fray Diego de Landa for the northern Yucatán (Tozzer 1941:174). The act of traversing the community or polity from its boundaries to its social centre would have engendered the same notions of social and spatial integration as the previous rituals. Finally, the base to summit of mountain processions act on the physical and supernatural levels to integrate not only the political and divine aspects of the ahau (Leventhal 1999), but to legitimate his role as a living embodiment of the axis mundi (Freidel 1992). One important aspect of this particular ritual activity is its integration with other aspects of directionality and architectural arrangement. For example, Reese-Taylor (2002:161) argues that this ritual activity may have involved the whole of epicentral Tikal, beginning at the Central acropolis and ending at the North acropolis. During this procession, the ritual actors would travel from a (relatively) low-lying watery locale, indicative of Xibalba (the Central Acropolis), through ballcourts, the portal to the Otherworld, and eventually to the North Acropolis, a series of raised temples linked directionally and architecturally to the heavens. More than any other example, Reese-Taylor (2002:162) notes that this particular ritual, as set within the architectural space of Tikal, is the "quintessential expression of the Mesoamerican north-south template".

The architectural elements which seem exclusive to Calakmul affiliates are amenable to such ritual processions. Calakmul paradigmatic element 1, which replicates that city's epicentre, has all of the physical requirements for such rituals, especially as found at Xunantunich and Naranjo, both of whose sacbeob would have permitted the periphery to centre processions just described.

The replication of Structure 2 at Calakmul affiliates might very well have been the setting for base-to-summit of mountain rituals, involving the role of the aligned polity within the Calakmul sphere of influence. If, as Richard Leventhal (1999) contends, the stairway ascending Xunantunich's El Castillo was designed with an eye for transformative ritual, could this not also entail the ritual and political patronage of Naranjo or Calakmul? Even if Xunantunich represents a breakaway state (Martin and Grube 2000:83) the structural similarities remain.

Likewise, Calakmul syntagmatic element 4 contains the necessary physical traits to accommodate ritual circuits. Although this is an inverted form of Tikal's southern residence, more examples in this study adhere to the Calakmul pattern. Ritual circumambulation, periphery to centre processions, and base-to-summit of mountain rituals could all have been accommodated through this architectural arrangement.

The claim that these common civic features represent ritual activity is based on the contentions that: 1) politics and ritual were highly integrated during the Late Classic; and 2) Calakmul actively pursued these as a means of political integration. The first point can be readily addressed. As indicated in Chapter 2, one of the enduring characteristics of Maya civilisation was the conflation of religious and political authority (Demarest 1992; Freidel and Schele 1988a, 1988b; Lucero 2003). Hieroglyphic and iconographic data treat royal and divine authority as a single issue (Looper 1995, 1999).

The rulers of Calakmul, no doubt with these same ideas in mind, actively and aggressively pursued a series of ritual activities during the Late Classic period (Martin and Grube 1995). In fact, it could be argued that the real strength of Calakmul lay not in its military reach (which was nevertheless the most impressive of its time), but through ritual sanctions. Calakmul participated in repeated, long-term, ritual activities with at least nine separate polities during this time.

Perhaps what we are seeing in the commonalities of the architecture of the Calakmul affiliates, then, is the result of a shared participation in a culturally- specific set of integrative rituals, ones designed to sanction activity and authority under the supervision of Calakmul. Of course, Calakmul did have recourse to military coercion, but even this was sanctioned and directed through ritual concerns and supernatural authority.

Interestingly, Tikal presents a much weaker case for the connection between political influence and architectural form. As mentioned, the only Tikal trait which may represent political structure is the opposing eastern and western pyramidal structures. As this is found in such disparate cities as Copán and Palenque, it is tempting to see this as a means of political integration on the part of Late Classic Tikal. Perhaps this was the case, but since these traits are the only example from the Tikal city plans which may indicate such a connection, the case cannot be made with confidence. Probably, the opposing pyramidal structures did form the physical component of such political rituals, but if this is so, we must assume that Tikal lagged far behind Calakmul in pursuing similar political connections, and in sponsoring or engendering the creation of their architectural symbols of power. In short, Calakmul, rather than Tikal, seems to indicate a correlation, although selective, between civic layout and political organisation, which more precisely relates to ritual activity.

City State Culture. This interpretation accords with the contention that the Classic period Maya were, indeed, a true city state culture (contra Braswell et al. 2004). According to Herman (2000:17), even during times of warfare and conflict "there is always considerable economic, religious, and cultural interaction, which crosses all frontiers" in a city-state culture. This would also account for the effective use of such symbols and rituals throughout the lowland Maya world, as they adhered to shared cultural aesthetics (Koontz and Reese-Taylor 2001).

However, what of secondary affiliates? In city state cultures, especially those that form hegemonic alliances (as we can assume the Calakmul superstate to have done), autonomy of client polities was both retained and conditional (Grube 2000; Iannone 2001). The disciplinary actions of Calakmul against its client Naranjo in 631 A.D. attest to such a situation (Martin and Grube 2000:72). While with primary affiliates we have written confirmation of hierarchical relations with the hegemonic polities, we remain unsure if secondary affiliates arranged their civic areas based on the ritual practices of Calakmul, or of Calakmul's primary affiliates. Based on the evidence covered in this study, it is impossible to say for certain at which level of emulation or inclusion the secondary affiliates were operating. However, the application of social memory and habitus may account for such architectural similarities.

As participants within the political sphere of Calakmul's dependents, secondary affiliates would have of necessity, required a "working knowledge" of the political and ideological subtleties of the relevant cities' administrative and diplomatic apparatus. One aspect of this "working knowledge" was an understanding of the physical settings through which such interactions would have been carried out and in which, presumably as dependents, the secondary affiliates would have taken part. These experiences would have been enshrined as social memories, and formed part of the corpus of images whereby the secondary affiliate's populace recognised and understood political authority. In short, the primary affiliates, like Calakmul itself, presented an image to the nobility of secondary affiliate cities of what authority looked like, an architectural and spatial metaphor which referenced ancient Maya understandings of power (Mosher 2004). It was this common reality (habitus), more than overt political control, which may account for secondary affiliate participation in the aforementioned architectural trends. If this was indeed the case, it is easy to see why Tikal, whose layout stands in stark contrast to most of the cities in this study, failed to assemble as effective a power network as did Calakmul, which relied on presumably more common architectural arrangements.

Part of the corpus of symbols and activities which secondary affiliates may have adopted from primary ones could have been the same integrative rituals described above. Again, such reproductive practices invoked the memory of historical authority, and can best be understood as an appropriation of the signs of political legitimacy, rather than a sign of political fealty. Without more secure hieroglyphic data on the role of secondary affiliates, however, this scenario remains tentative. While social memory may indeed account for such purposes, it does not necessarily mean that it is the case.

SUMMARY

Based on the relevant architectural markers of power, and their inferred ritual and integrative functions, I believe that Calakmul actively sponsored or influenced the creation of a series of ritually and politically-charged architectural types, at least as concerns its primary affiliates. What transpired with the secondary Calakmul affiliates is unclear, but a provisional interpretation that can be offered is that the social memory and familiarity of Calakmul's primary affiliates may have inspired or influenced the continuation of Calakmul architectural forms to secondary affiliates. Drawing on ethnohistoric and ethnographic information, as well as Classic period histories, a plausible explanation is that Calakmul supported its program of socio-political integration through a series of architecturally-specific rituals. These features are what account for the similarities between the various polities.

A less secure case can be made for the connection between Tikal civic elements and political structure. Only one feature displays such a connection (the opposing pyramidal structures), and even so, not nearly with the same numbers as Calakmul. As participants in a shared Classic Maya city state culture, we can assume that Tikal did indeed operate in similar ways to Calakmul in respect to socio-political integration, but either the cities studied do not adequately reflect the reality of Tikal's political structure or, conversely, the hieroglyphic record and the architectural record are mutually supportive. In both sets of data, Calakmul demonstrates a more vigorous, and larger, socio-political hegemonic state than does Tikal during the Late Classic period. In short, we can conclude that although the relationship between architecture and socio-political structure is much more complicated than a simple matter of correlation, there does seem to be, at least in the case of Calakmul, some merit to Ashmore and Sabloff's (2002) initial claim that civic plans embed information pertinent to understanding interpolity hierarchies.

CHAPTER 6

CONCLUSIONS

At the outset of this study, I presented several questions to frame the research into ancient Maya civic plans. In order to assess these questions as guiding factors to this exercise, they are addressed individually below:

1. What are the Definitive Architectural Factors which Separate the Tikal and Calakmul Layouts?

As mentioned in Chapter 4, the separate architectural elements which differentiate the Tikal and Calakmul civic templates are to be found in the paradigmatic and syntagmatic elements which constitute such plans. Briefly, these elements are as follow:

Tikal Paradigmatic: 1. Main north-south axis in site organization; 2. Twin Pyramid Groups; 3. Formal and functional dualism between the north and south; 4. The addition of elements on the east and west to form a triangle with the north.

Tikal Syntagmatic: 1. Medially-placed ballcourt; 2. North-situated, south-facing pyramidal structures; 3. Eastern and western situated pyramidal structures; 4. Southern-situated acropolises; 5. Radially-constructed pyramidal structures; 6. Northern, roofless, Twin Pyramid Group structure; 7. Internal sacbeob.

Calakmul Paradigmatic: 1. Epicentral layout reproduction; 2. North-south emphasis on site organization.

Calakmul Syntagmatic: 1. Structure 2-like building; 2. North-south plaza with medial structure; 3. Main plaza framed by an E-Group; 4. A northern-situated residence raised above the main plaza.

Obviously, some of these features are cross references of each other (for example, Tikal Paradigmatic Element 1 and Calakmul Paradigmatic Element 2). Although the specific elements employed in all of the cities are of a few basic architectural forms (i.e., range structures, pyramidal platforms, temples, etc.), these syntagmatic elements were consciously reworked to form paradigmatically distinct civic templates, which, I have argued, were possibly related to differing integrative ritual practices.

2. What Specific Cosmological, Ideological and Experiential Aspects of ancient Maya Society are Alluded to in these Templates?

Ancient Maya perceptions of time, directionality, the axis mundi, and the multilayered universe all factor into the meanings of individual structures, and, by extension, civic plans. Appropriate to this conversation is the notion of cosmograms, which created the spatial references by which the ancient Maya organized their cities. Appropriately, most of the information that has come down to us from the Classic period Maya pertaining to these issues involve the conjoined ideas of religion and kingship. Through the creation of specific architectural forms and civic plans, the ancient Maya were manifesting and expressing a shared belief in the conflation of cosmology and political authority. On a more mundane level, the commonality of the built environment, with or without political or ritual prestige, may well have been ingrained as social memory, and hence guided architectural projects not so much owing to cosmological or political ideals, but rather, as the image of what authority "looked like".

3. Does the Emulation of these Templates Imply Appropriation of Political and Religious Ideals or Membership in a Superstate?; also,
4. What was the Operational level of emulation? Were all Considered Polities Actively Involved in Fashioning their Respective Centres After Tikal or Calakmul, or was there a More Dispersed Pattern, where Secondary and Tertiary Centres Arranged Themselves After Allied Primary Centres which Aimed to Reproduce the Seat of the Tikal and Calakmul Polities?

These questions, although not known at the outset of this study, can be answered in unison. I would argue, based on the evidence covered in this study, that the reproduction of specific elements amongst primary affiliates, especially those of Calakmul, may indicate membership in a ritually-bound political structure. Of course, the juxtaposition of religion and political authority is a false one, insofar as it applies to the ancient Maya. In terms of secondary affiliate polities the situation is much less clear, and the operational level of emulation or inclusion remains in doubt. Quite possibly, these centres were also directly tied to the ritual and political concerns of Calakmul the primary affiliates. In such a scenario, the closest that we can come to explaining the continuation of these architectural forms, is that the primary affiliates affected the secondary ones in the same manner as Calakmul did the primary ones, thus instituting the physical and spatial expressions of power to which they themselves were familiar.

5. Does the Civic Plan Evidence Correlate with the Epigraphically Reconstructed Political History?

As with much of this study, the answer to this question is both yes and no. There does seem to be a stronger correlation between Calakmul's civic elements and its epigraphically-designated political allies. That being said, only one paradigmatic element (1), and two syntagmatic elements (1 and 4) can be said to support this. Tikal has a possible connection between political history and architectural form in its syntagmatic element 3. However, as stated earlier, this is a tenuous assertion. The majority of the similarities between architecture and political history seem restricted to ritual activities which, as we have seen, were as important as any part of ancient Maya politics.

Interestingly, the pattern of shared architectural elements (as cultural aesthetics) reifies the interpretation that the syntagmatic meanings that have been derived in this study are seeing in the civic plans of the ancient Maya, is a facet of a city state culture. Like economics, the ritual activity and cosmological imagery associated with the various civic elements had a wide distribution, and were amenable to particular socio-political contexts, as well as to supporting the larger frameworks of the ancient Maya, such as macro-political superstates. This, coupled with the probable ritual function of the selected civic elements, makes possible the argument that there was a correlation between ancient Maya superstates and selected architecture, but only in very rare cases did this approach the entire paradigmatic "meanings" of the main polities themselves (as in the case of Naranjo).

However, the issue of autonomy, especially in regard to secondary affiliates, is a serious point to consider in ascribing political influence to those centres which were politically aligned and which participated in similar architectural traditions. Local autonomy, a central tenet of city state cultures, does not always harmonise with external political influence, as was the case in the previously mentioned example of the annexation by Calakmul of its dependency, Naranjo's ruling house (Chapter 2). Rather, what we are seeing in the correlation of political history and civic architectural elements is more likely the infrastructure for a shared ritual relationship (paradigmatic). The syntagmatic "chains of evidence" (Iannone 1993:16) implicate specific ritual activities which, I postulate, were utilised as a political adhesive – affirming and reifying the paradigmatic unity of Calakmul's hegemony, couched in familiar, architectural cosmological terms. This was a significant means of socio-political integration, nevertheless, as the epigraphic record of Calakmul attests.

However, we should be wary of equating ritual relations with political overlordship. Such a statement goes beyond the data covered in the present study, and must be investigated on its own terms. Rather, we should take the convergence of epigraphic and architectural data as a beginning point from which to explore issues of political relations, especially political autonomy.

SUMMARY

In sum, while this study has generated some support for Ashmore and Sabloff's (2002) proposition, it is of a very selected nature. There does, indeed, seem to be a correlation between political structure and civic arrangement. However, this relation rests on a few key traits, specifically those amenable to ritual circuits, presumably of a politically-integrative nature. Also, the degree to which secondary polities participated in such structures, as Martin and Grube's (1995) "superstate" proposition, needs to be adequately addressed by more thorough examinations of the interplay between civic layout, political histories, and archaeological excavations, with an eye to determining the limits and constraints of hegemonic city state systems. That being said, this study of the civic plans of ancient Maya polities does indicate that they hold largely untapped information relating to ancient Maya society, including politics, cosmology, and social structure. Hopefully, more research will be generated on this topic, and the questions remaining from this and other studies on the interaction between ancient Maya ideology, society, and the built environment will be addressed.

TABLE 1. PRIMARY POLITICAL AFFILIATES OF TIKAL AND CALAKMUL
(after Martin and Grube 2000).

Tikal	Calakmul	Ambigious
Palenque	Dos Pilas	Yaxchilan
Motul de San Jose	Naranjo	Tonina
Ixlu	El Peru	Pomona
Jimbal	Caracol	
Uaxactun	Cancuen	
Yaxha	Moral	
Copan	Piedras Negras	
Maasal	Maasal	
	Quirigua	
	Oxpemul	
	Nadzcaan	
	La Muñeca	
	Ixkun	
	Pomoy(?)	
	La Corona	

TABLE 2. SECONDARY POLITICAL AFFILIATES OF TIKAL AND CALAKMUL
(after Martin and Grube 2000).

Tikal	Calakmul
Toktan	Arroyo de Piedra
Oxte'kuh	Itzan
Yaxha'	Yaxha'
Anayte	Tamarandito
Comalcalco	Aguateca
Pusilha	Aguas Calientas
Nim Li Punit	El Caribe
La Mar	La Mar
Santa Rita	La Rejolla
Xkuy	Ixkum
	El Chorro
	Holmul
	Xunantunich
	Ucanal
	Yootz
	Kina
	El Cayo
	La Amelia
	Cahal Pichik
	Ko-Bent-Cauc
	Tzam
	Hix Witz
	Sak T'zi
	Chapayal
	Ahkul
	Yaxha

TABLE 3. RITUAL ACTIVITY AMONGST POLITIES

POLITY	RITUAL PARTNER
Tikal	Maasal/Palenque/Copan?
Palenque	Tikal
Maasal	Tikal
Copan	Quirigua
Dos Pilas	Calakmul
Naranjo	Ucanal/Yootz
El Peru	Calakmul
Caracol	Tikal/Calakmul/Ixkun/Ucanal
Cancuen	Dos Pilas/Calakmul
Moral	Calakmul
Piedras Negras	Calakmul/La Mar/El Cayo
La Corona	Calakmul
Quirigua	Copan/Calakmul
Calakmul	Caracol/Naranjo/Dos Pilas/Piedras Negras/Moral/El Peru/Cancuen/La Corona
Tamarandito	Dos Pilas
Aguateca	La Amelia
Yaxchilan	Bonampak/Lacanah
Ucanal	Caracol/Naranjo
Yootz	Naranjo
Bonampak/Lacanah	Yaxchilan
Sak T'zi'	El Cayo
El Cayo	Piedras Negras/Sak T'zi'
La Mar	Calakmul
Xkuy	Quirigua
La Amelia	Aguateca

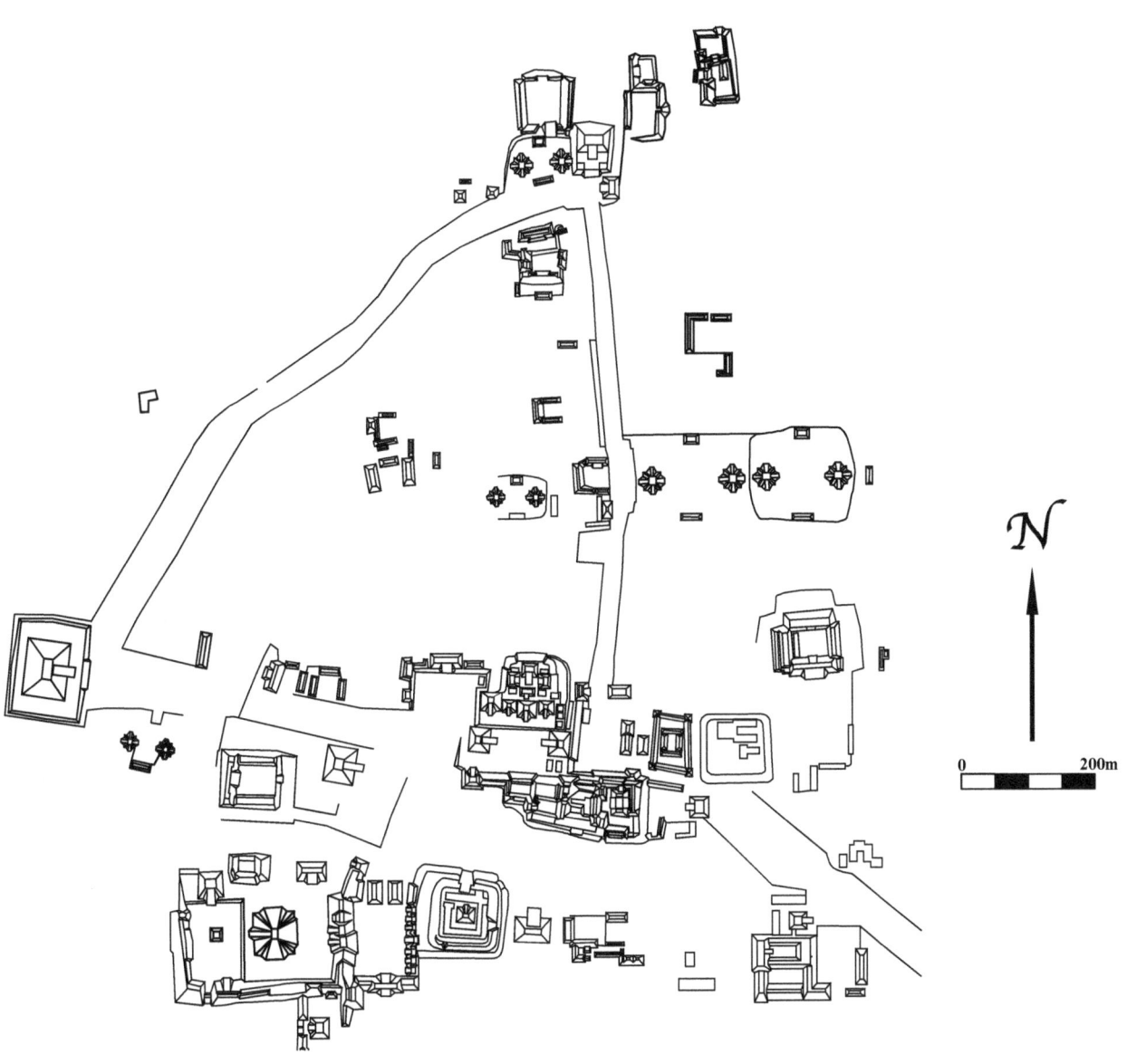

FIGURE 1. Map of Tikal, Guatemala (after Martin and Grube 2000:24).

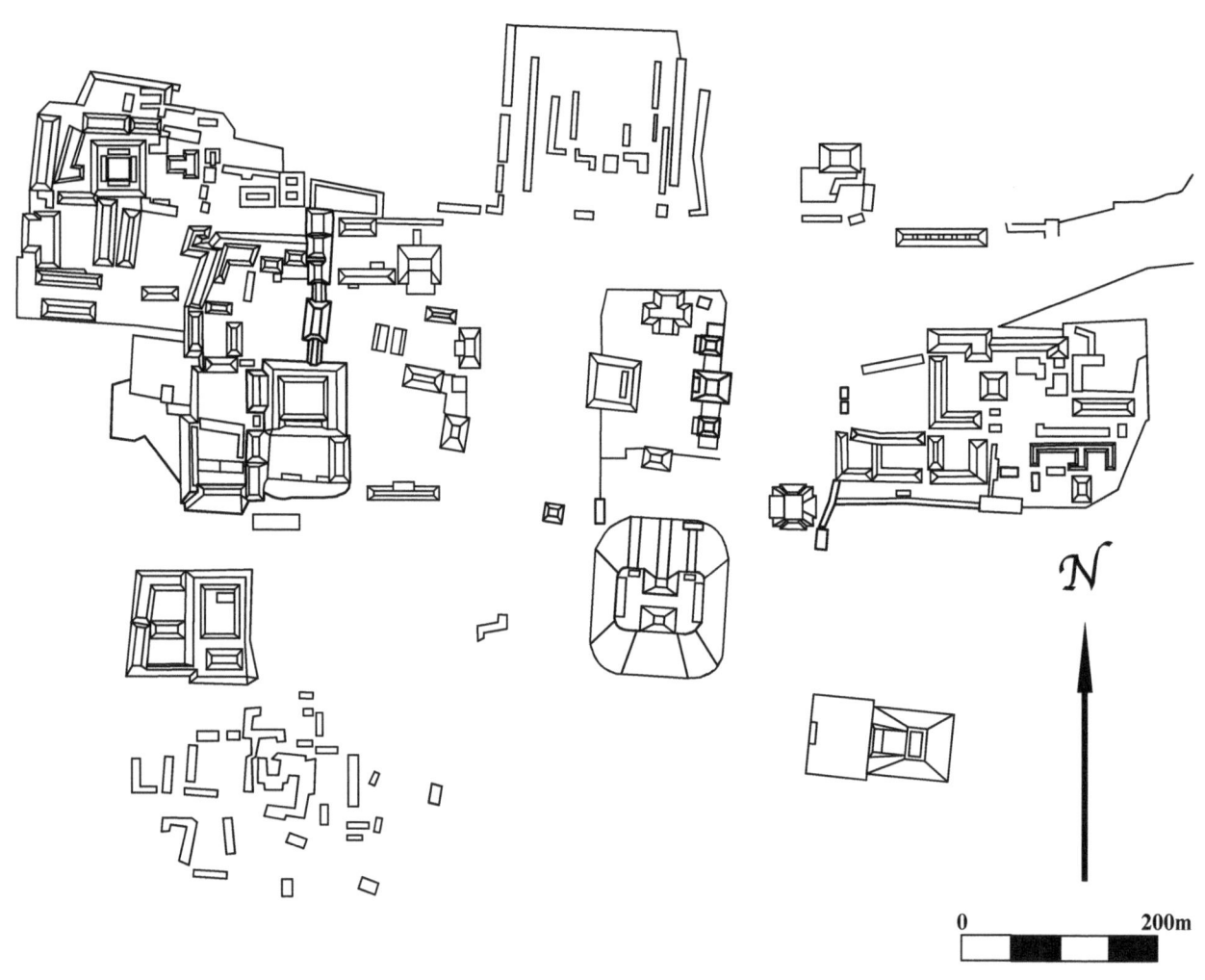

FIGURE 2. Map of Calakmul, Mexico (after Martin and Grube 2000:100).

FIGURE 3. Map of Aguateca, Guatemala (after Sharer and Traxler 2006:410).

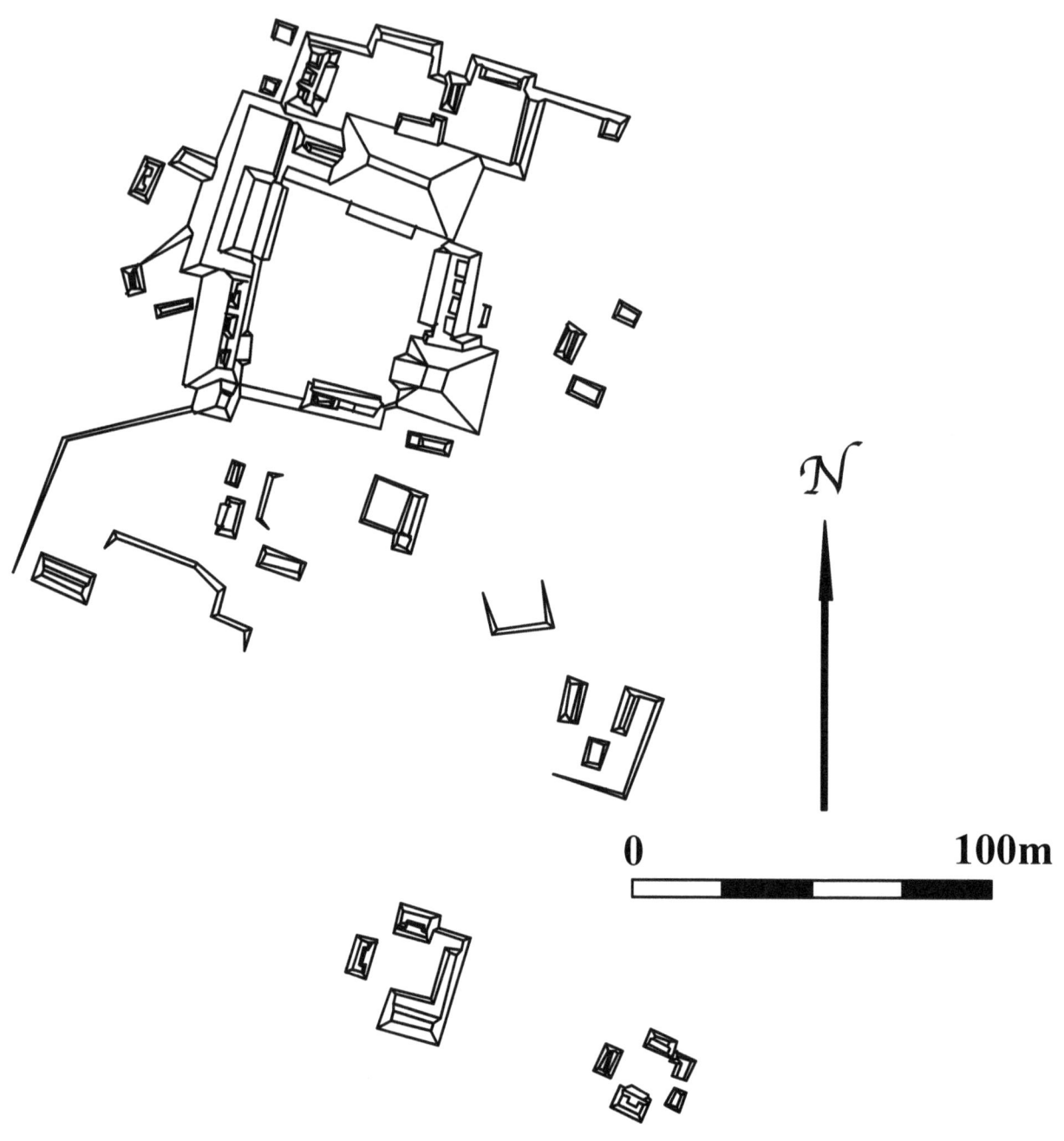

FIGURE 4. Map of Arroyo de Piedra, Guatemala (after Escobedo 1997:308).

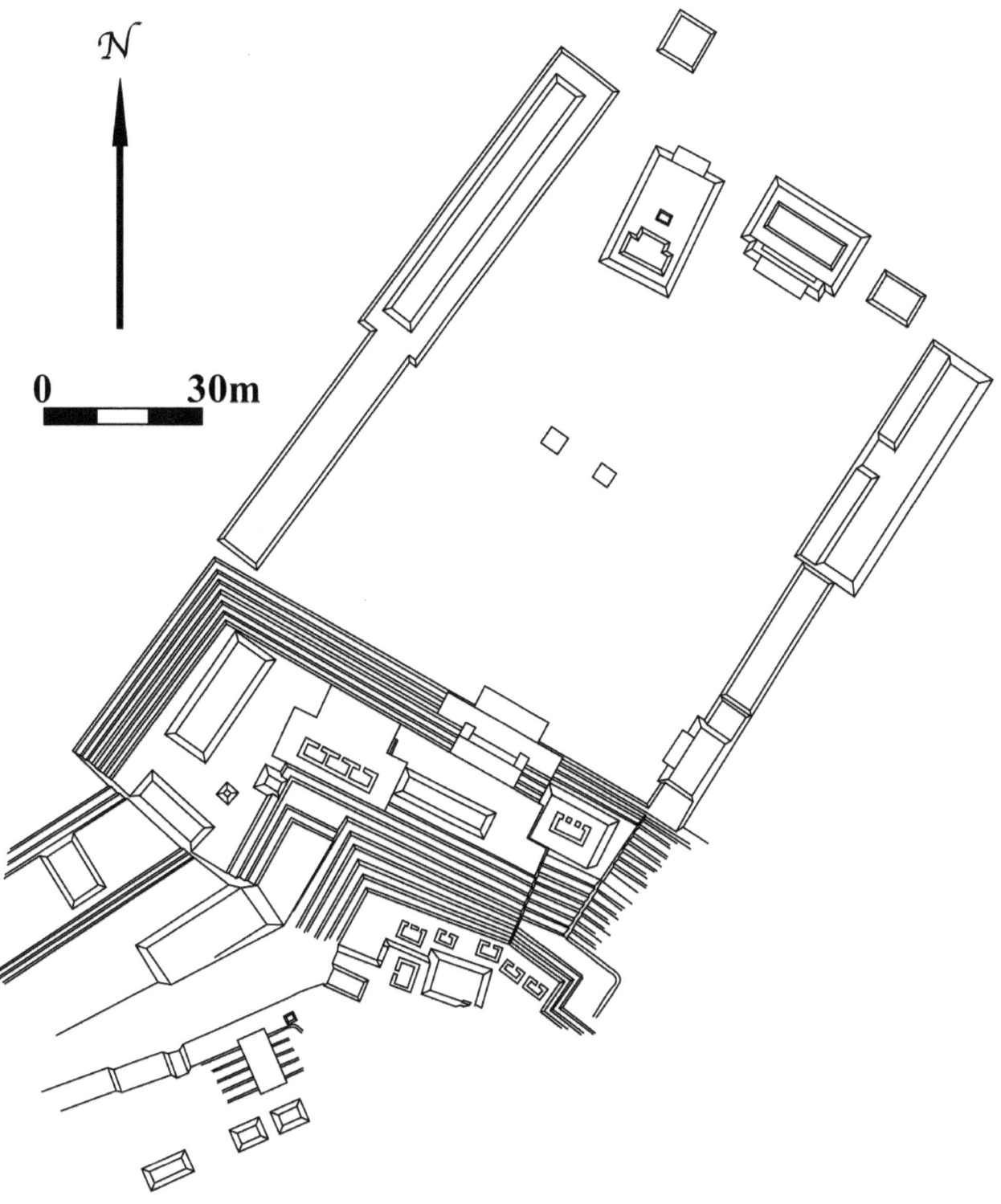

FIGURE 5. Map of Bonampak, Mexico (after Ruppert et al 1955).

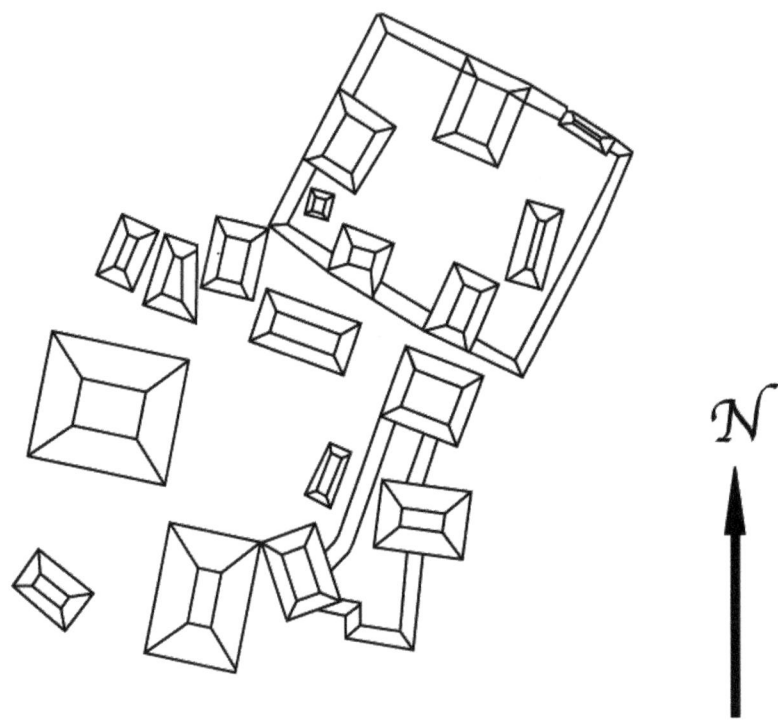

FIGURE 6. Map of Cahal Pichik, Belize (after Thompson 1931:240).

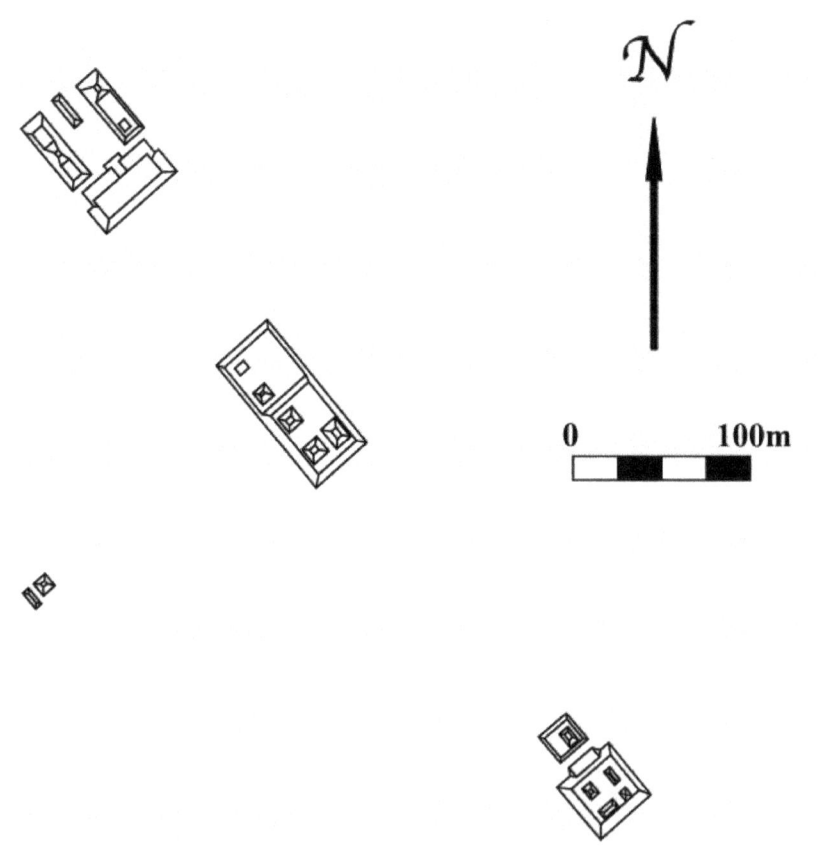

FIGURE 7. Map of El Cayo, Mexico (after Lee and Hayden 1988:12).

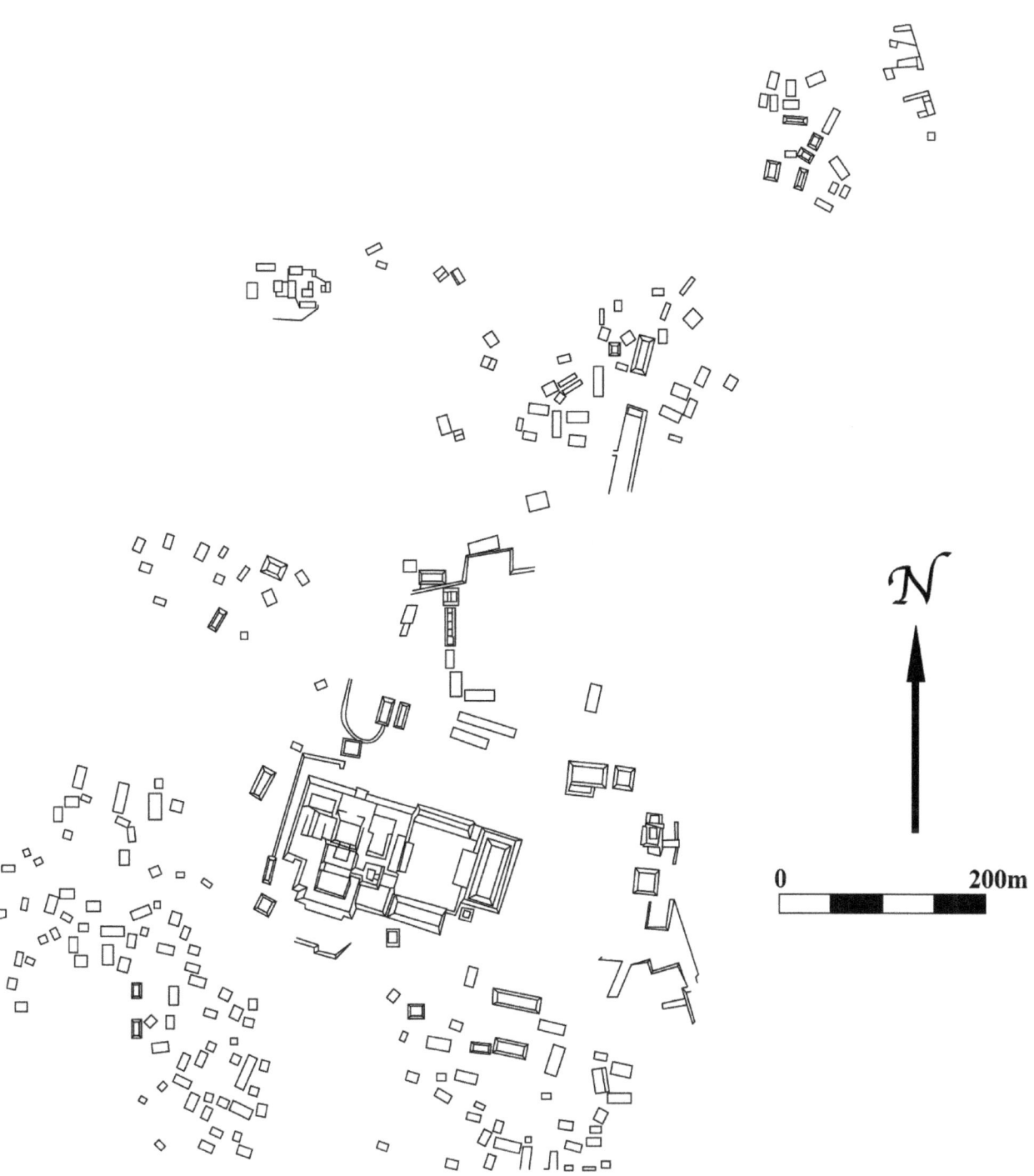

FIGURE 8. Map of Cancuen, Guatemala (after Cook et al 2005:631).

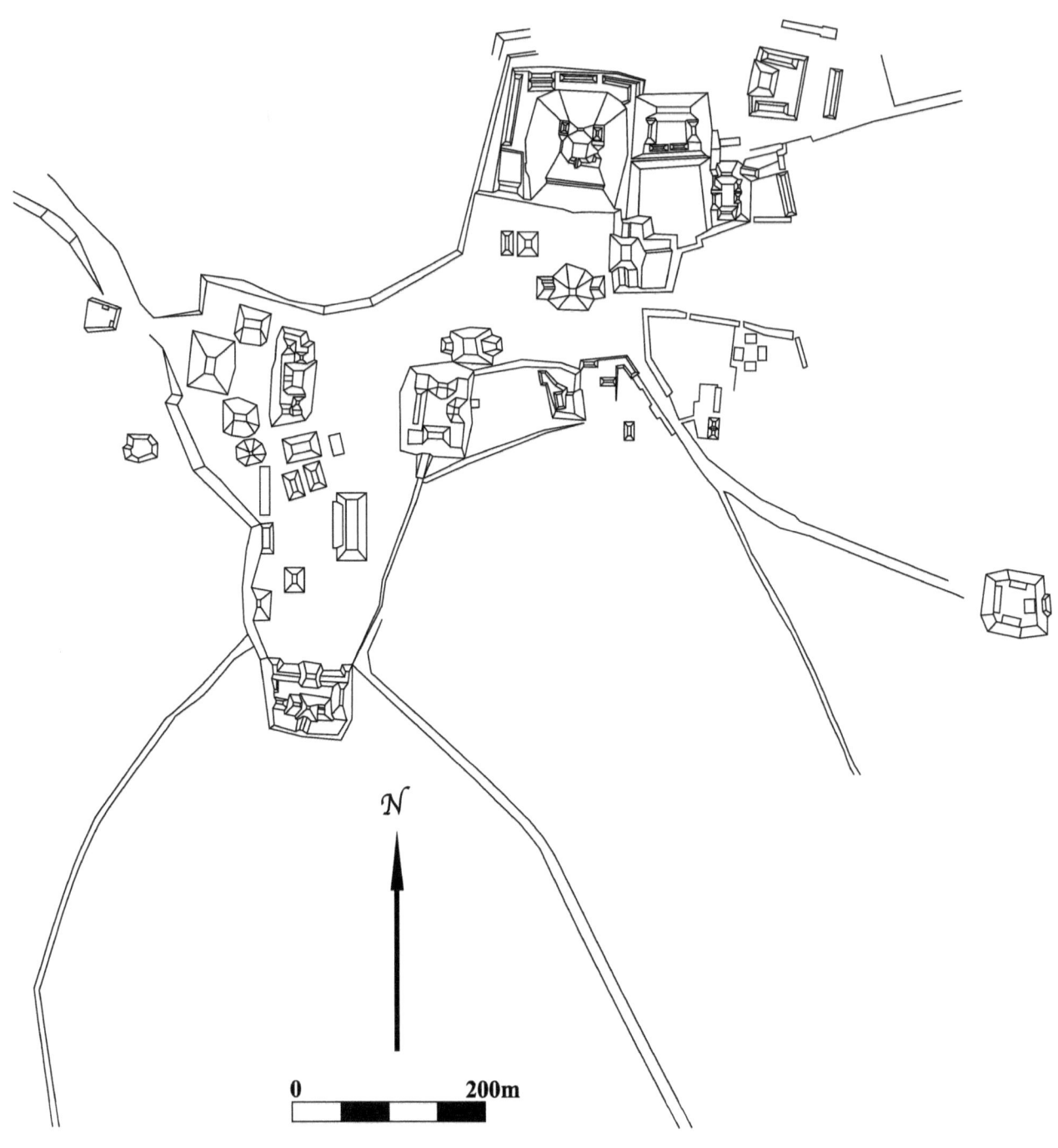

FIGURE 9: Map of Caracol, Belize (after Martin and Grube 2000:84).

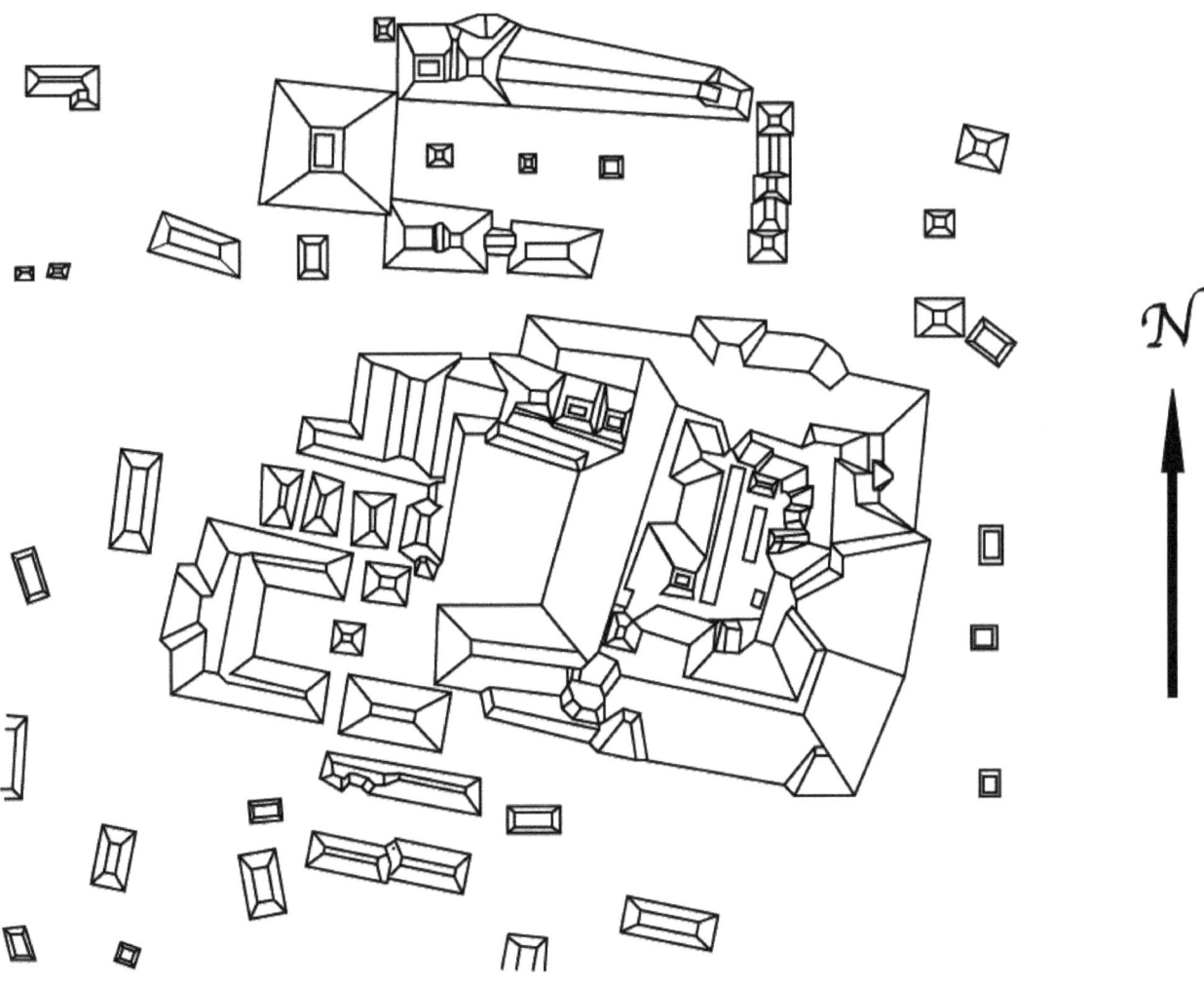

FIGURE 10: Map of Comalcalco, Mexico (after Andrews 2001).

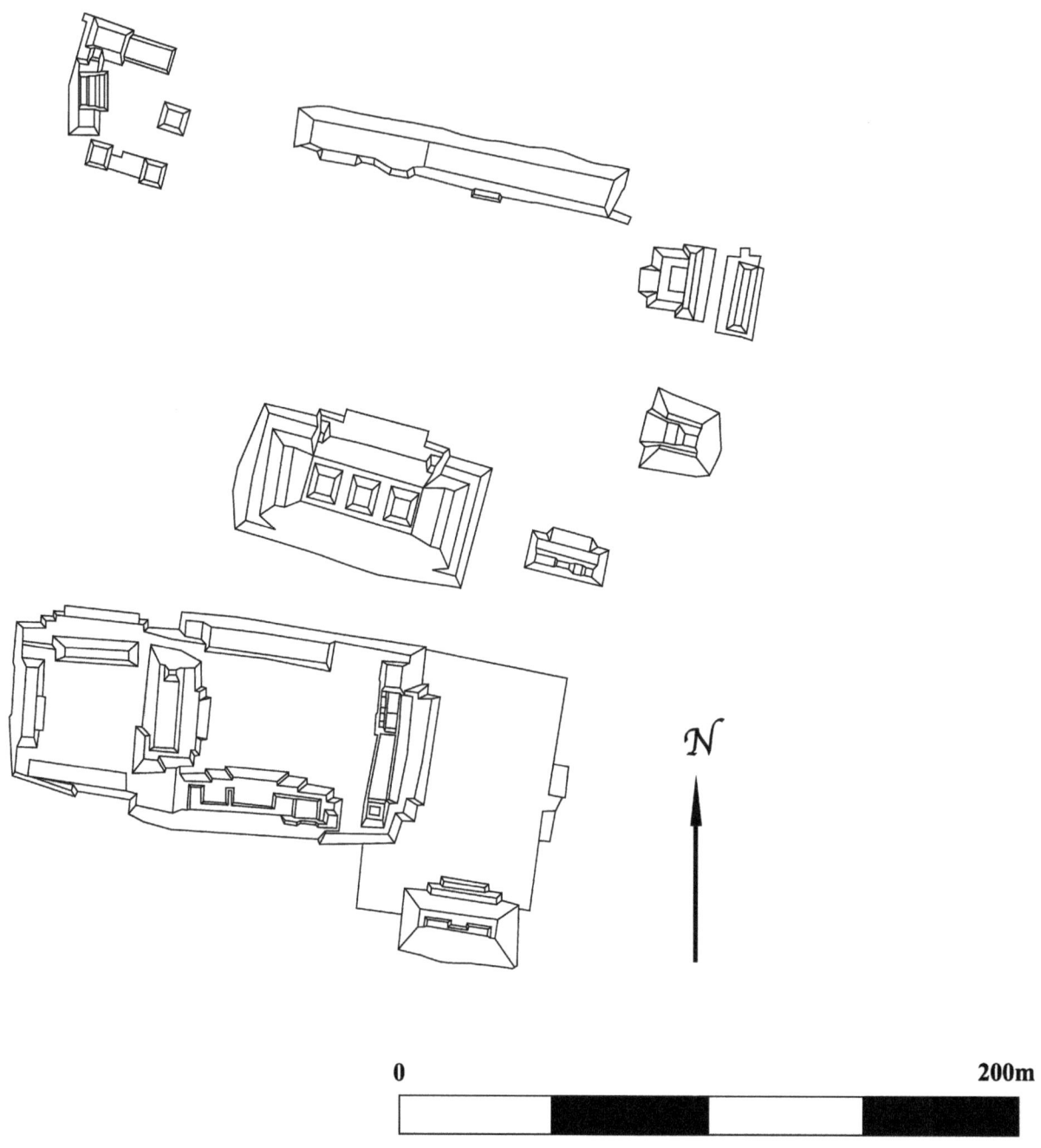

FIGURE 11. Map of Dos Pilas, Guatemala (after Martin and Grube 2000:54).

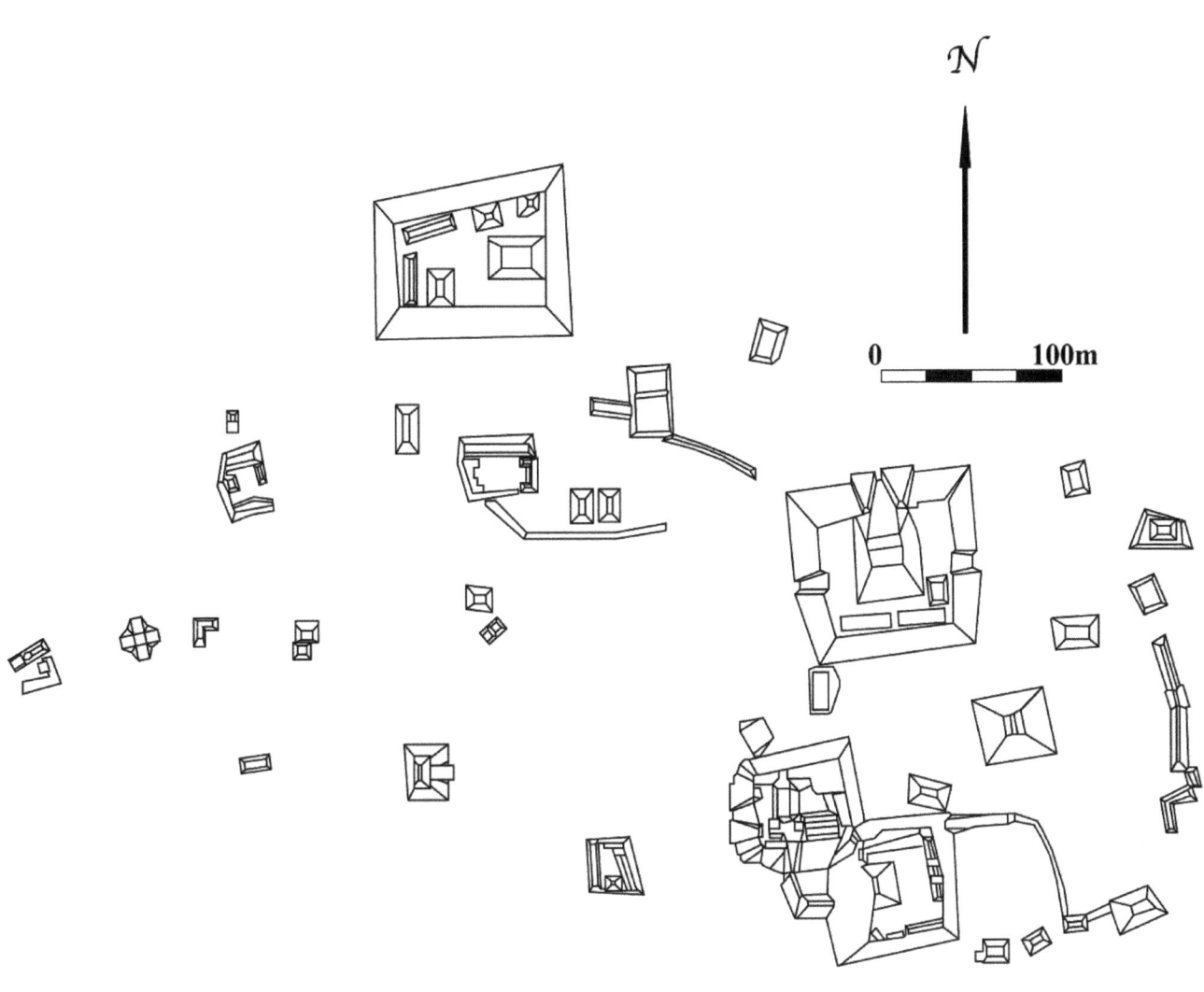

FIGURE 12. Map of Holmul, Guatemala (after Estradi-Belli 2000).

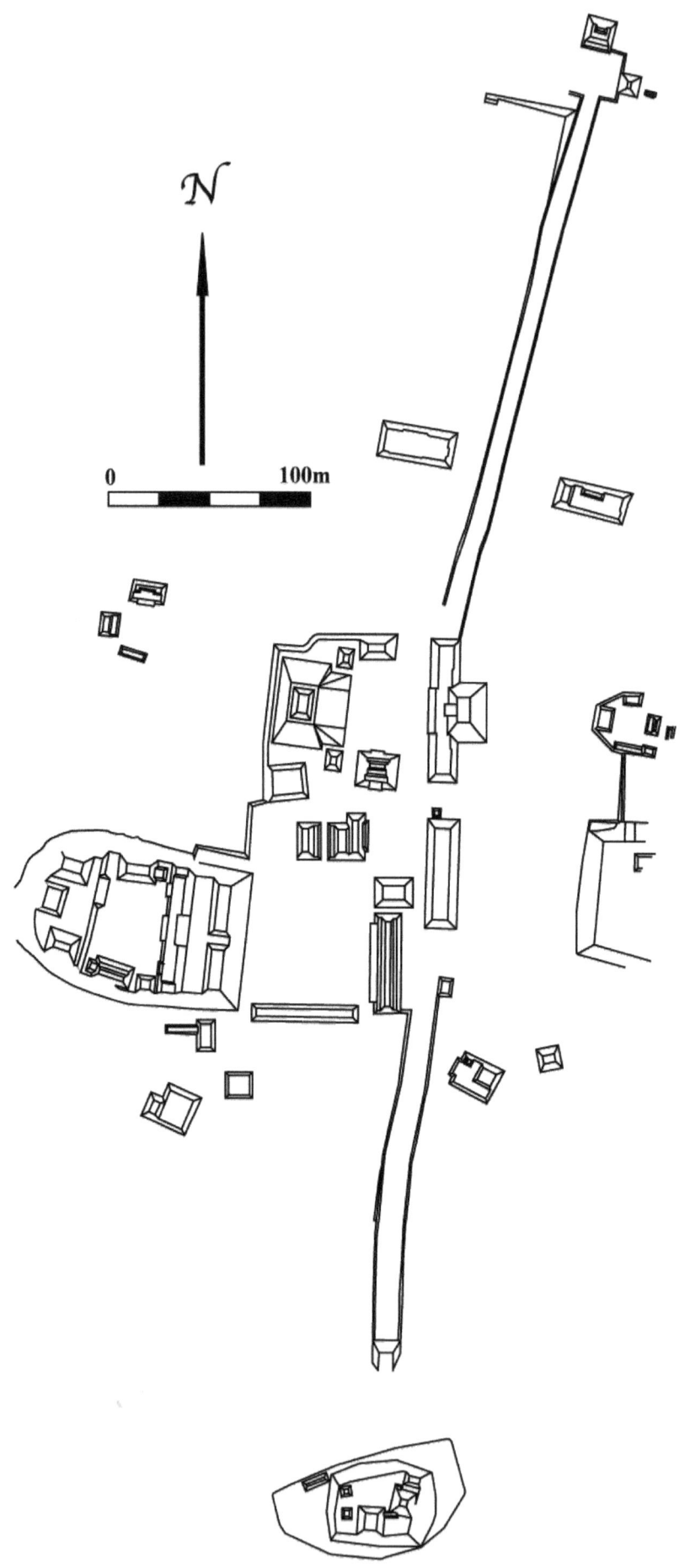

FIGURE 13. Map of Ixkun, Guatemala (after Graham 1980).

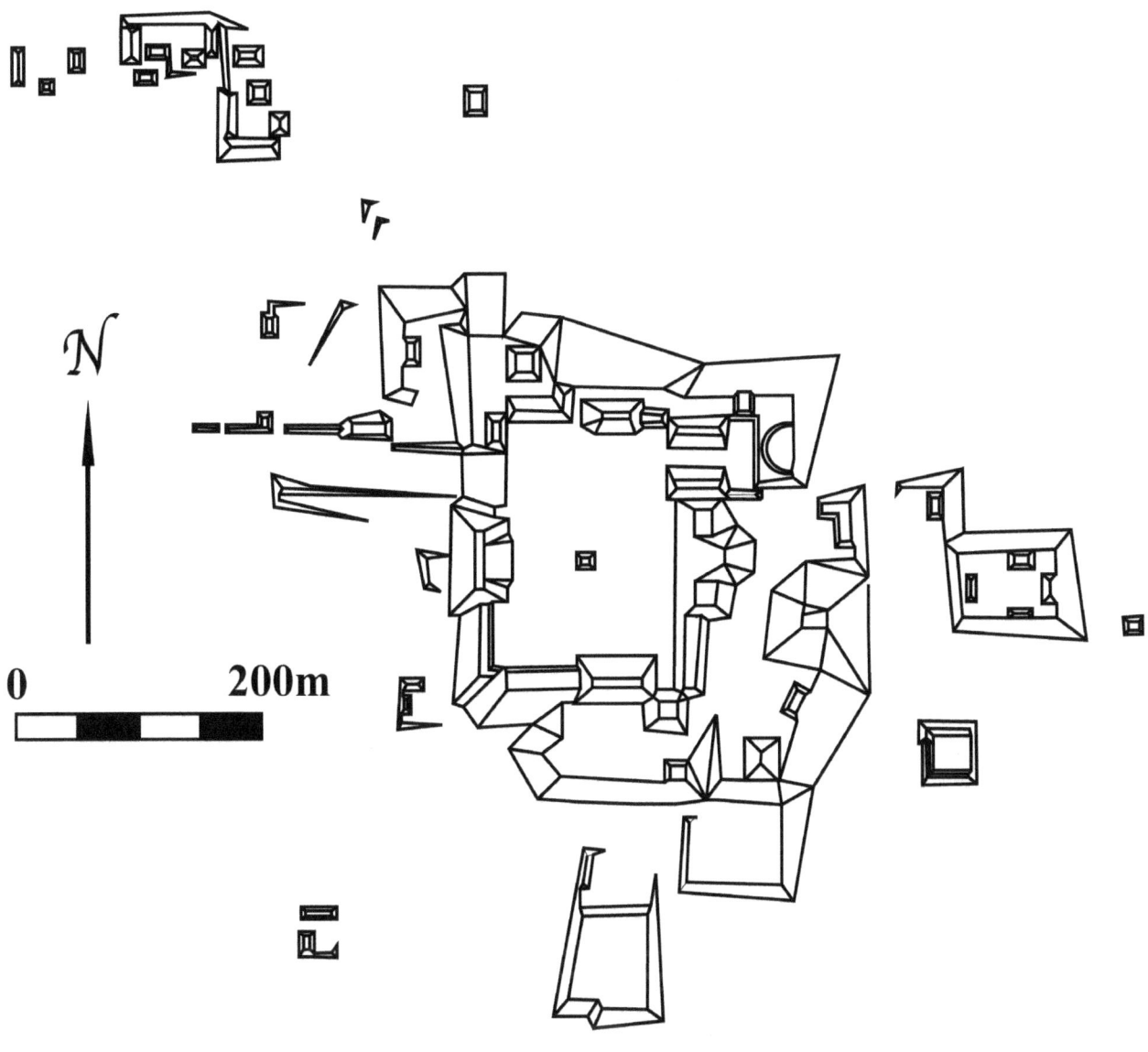

FIGURE 14. Map of Jimbal, Guatemala (after Puleston 1983).

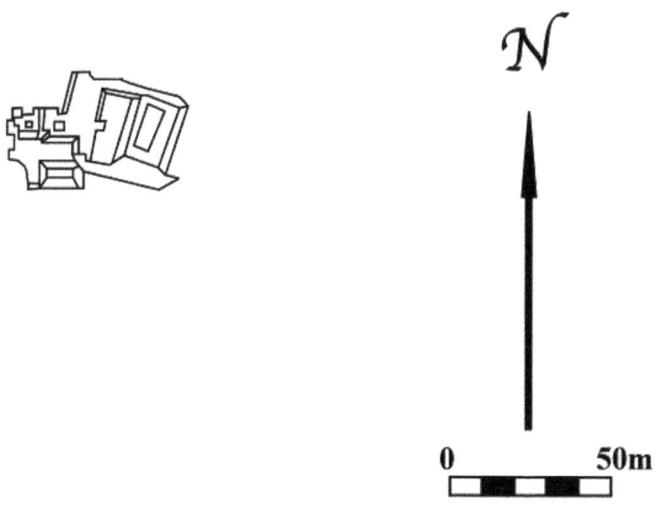

FIGURE 15. Map of La Pasadita, Guatemala (after Golden 1999).

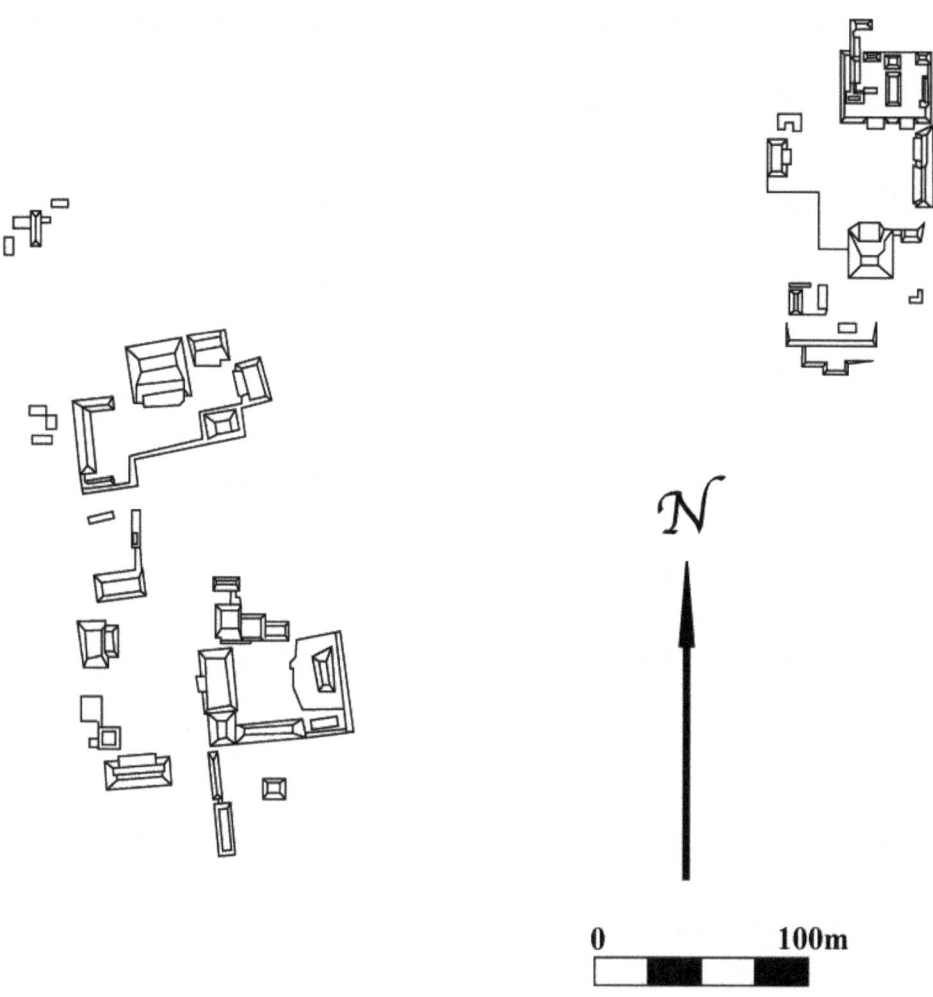

FIGURE 16. Map of Tamarandito, Guatemala (Valdés 1997:322).

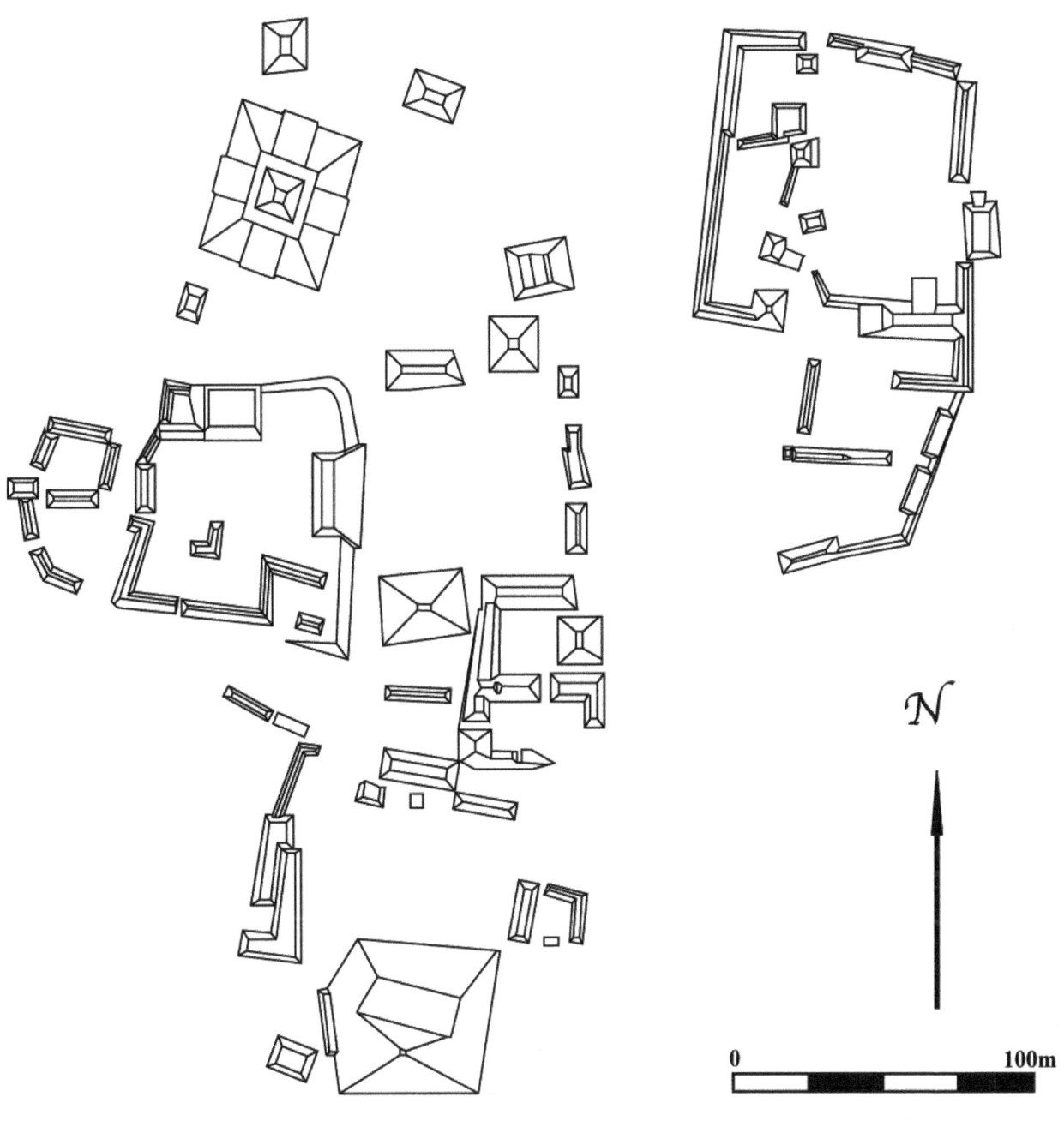

FIGURE 17. Map of La Muñeca, Mexico (after Ruppert and Denison 1943:10).

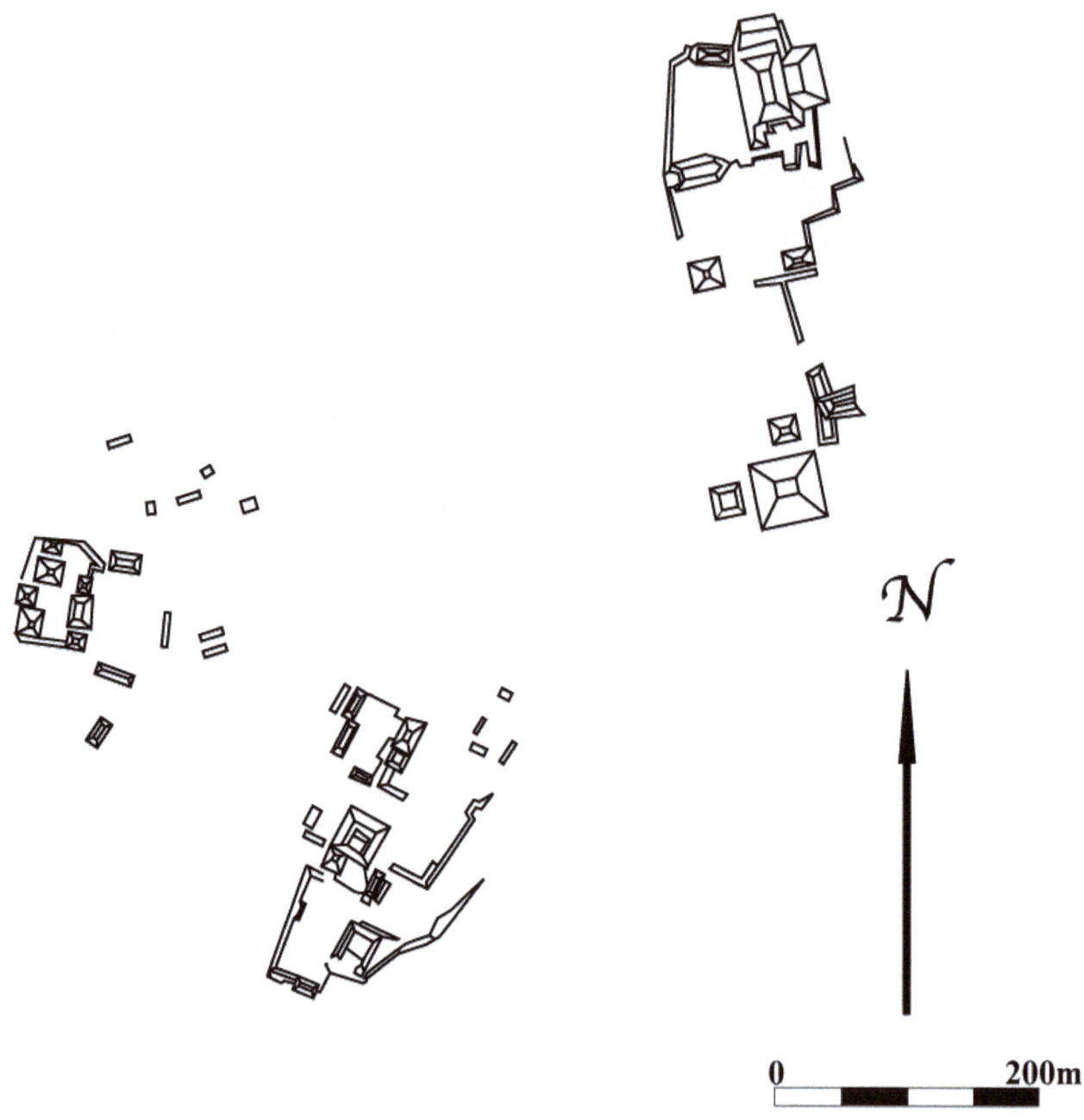

FIGURE 18. Map of Nadzcaan, Mexico (after Carrasco and Wolf 1996:70).

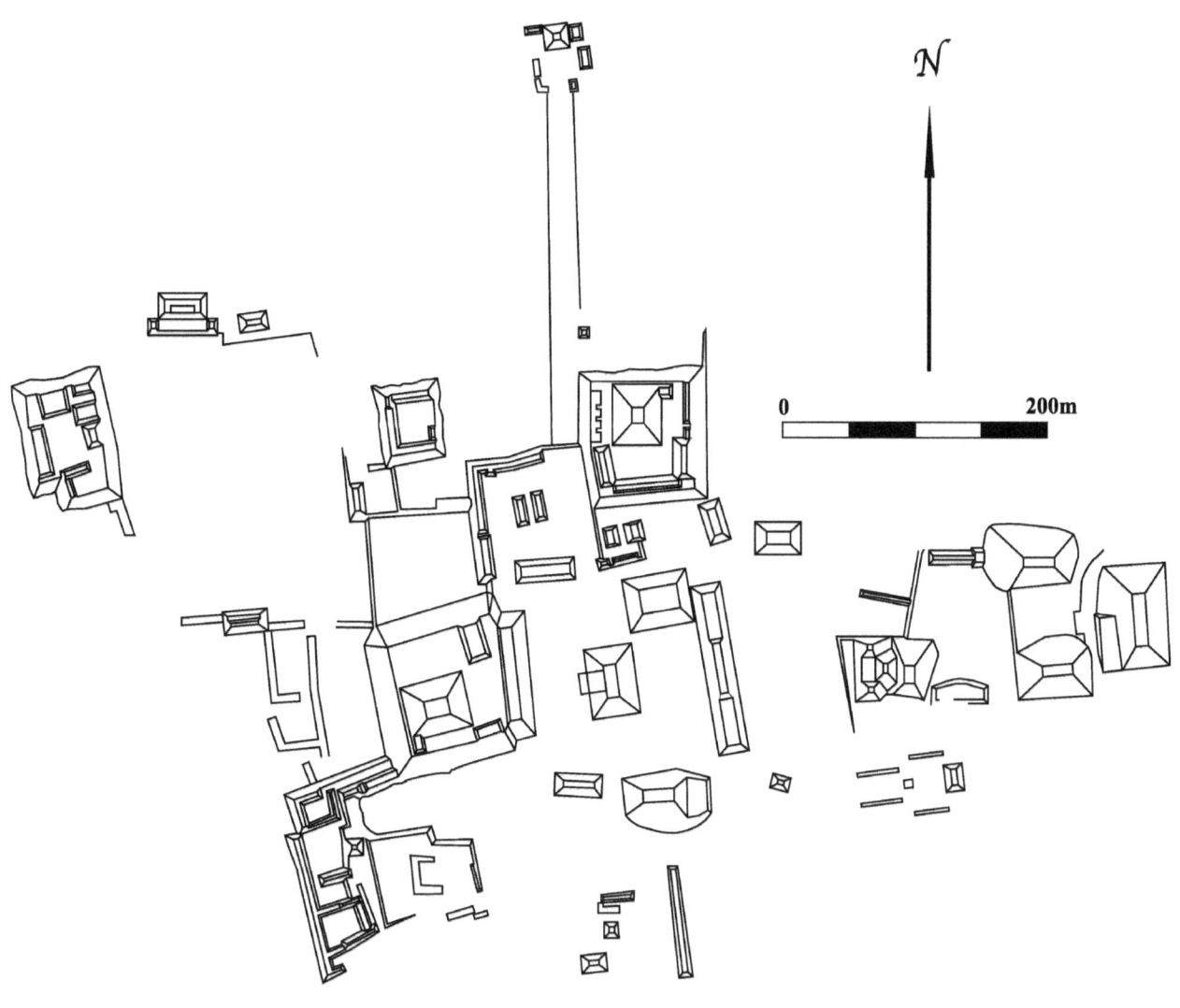

FIGURE 19. Map of Naranjo, Guatemala (after Martin and Grube 2000:68).

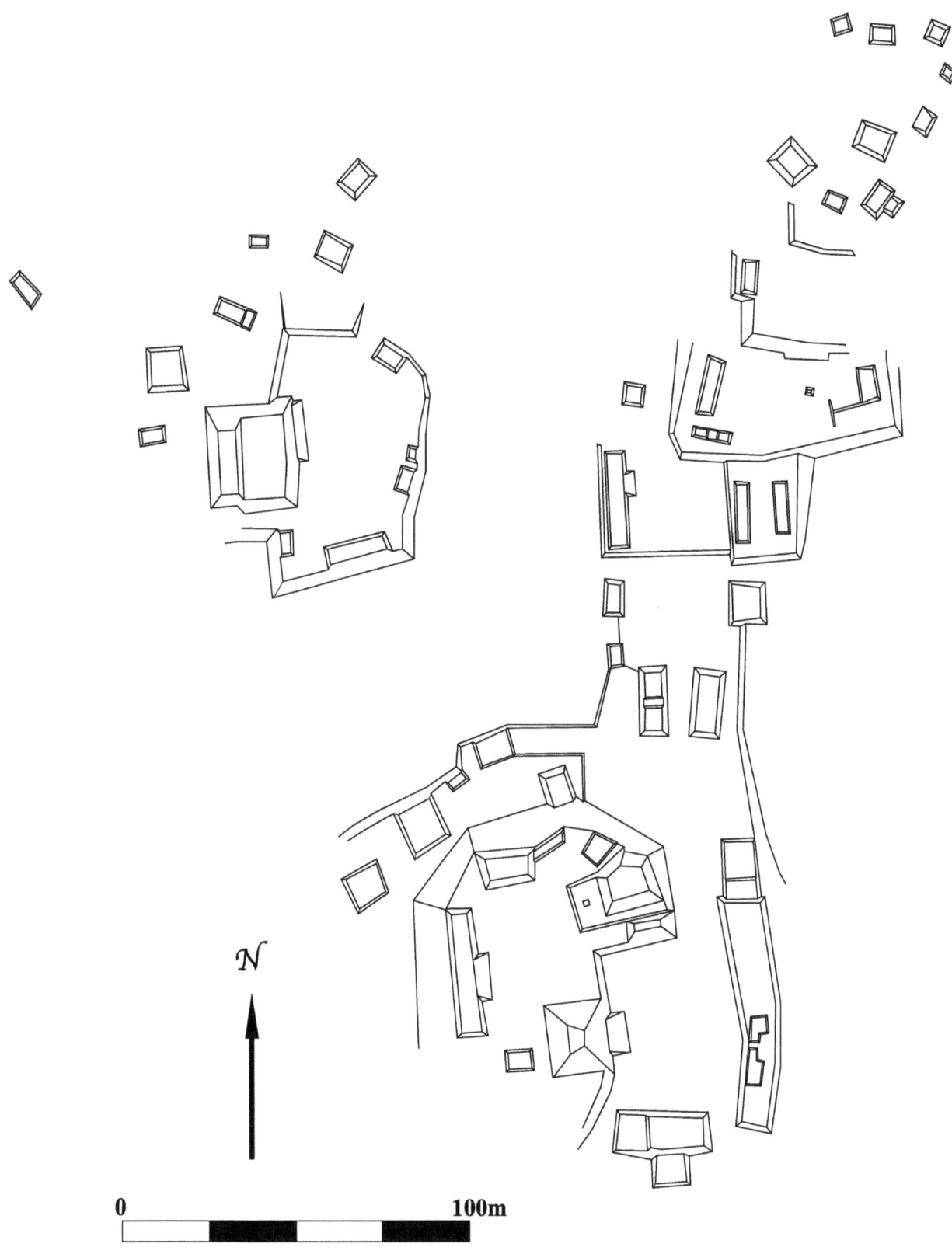

FIGURE 20. Map of Nim Li Punit (after Leventhal 1992:133).

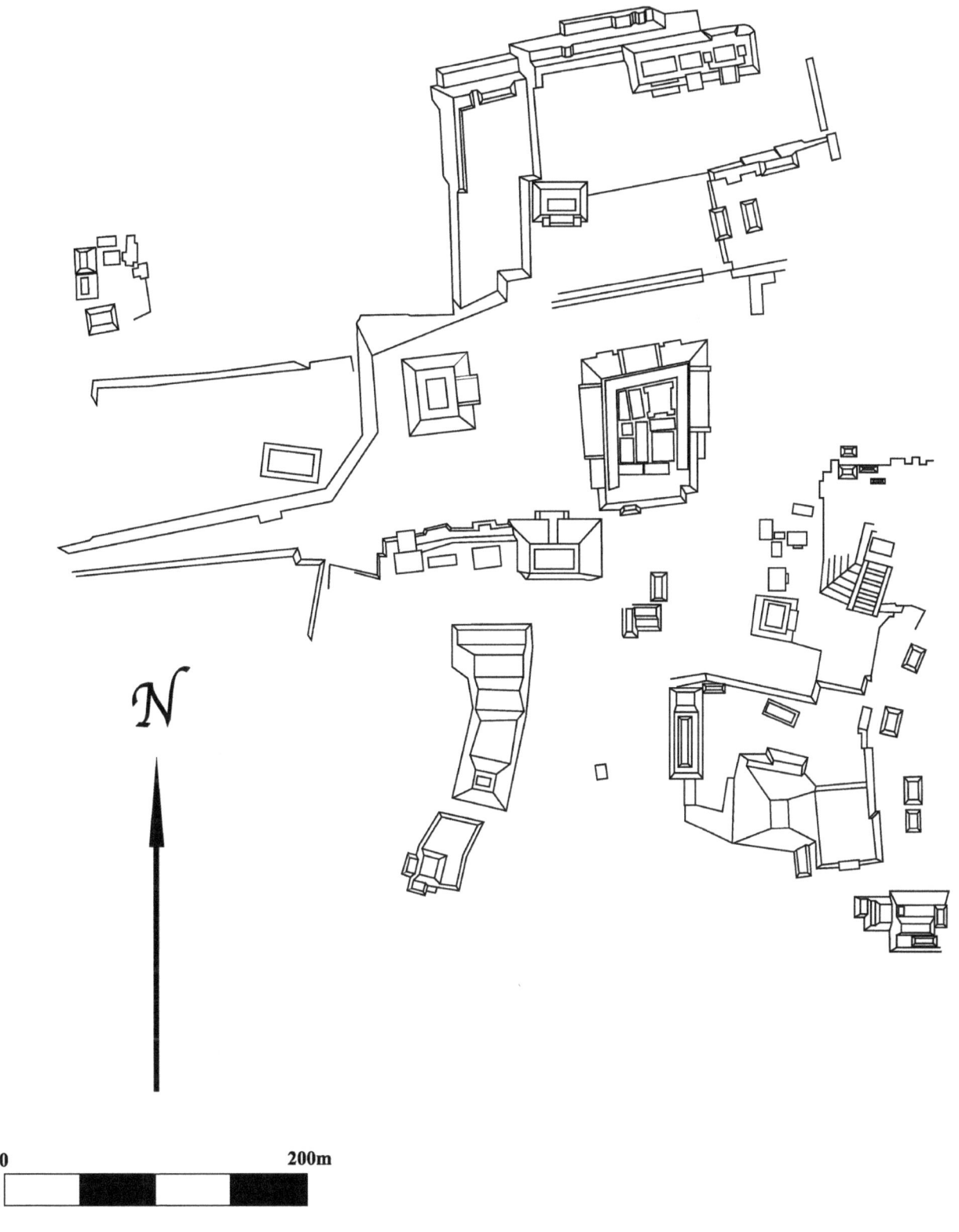

FIGURE 21. Map of Palenque, Mexico (after Martin and Grube 2000:154).

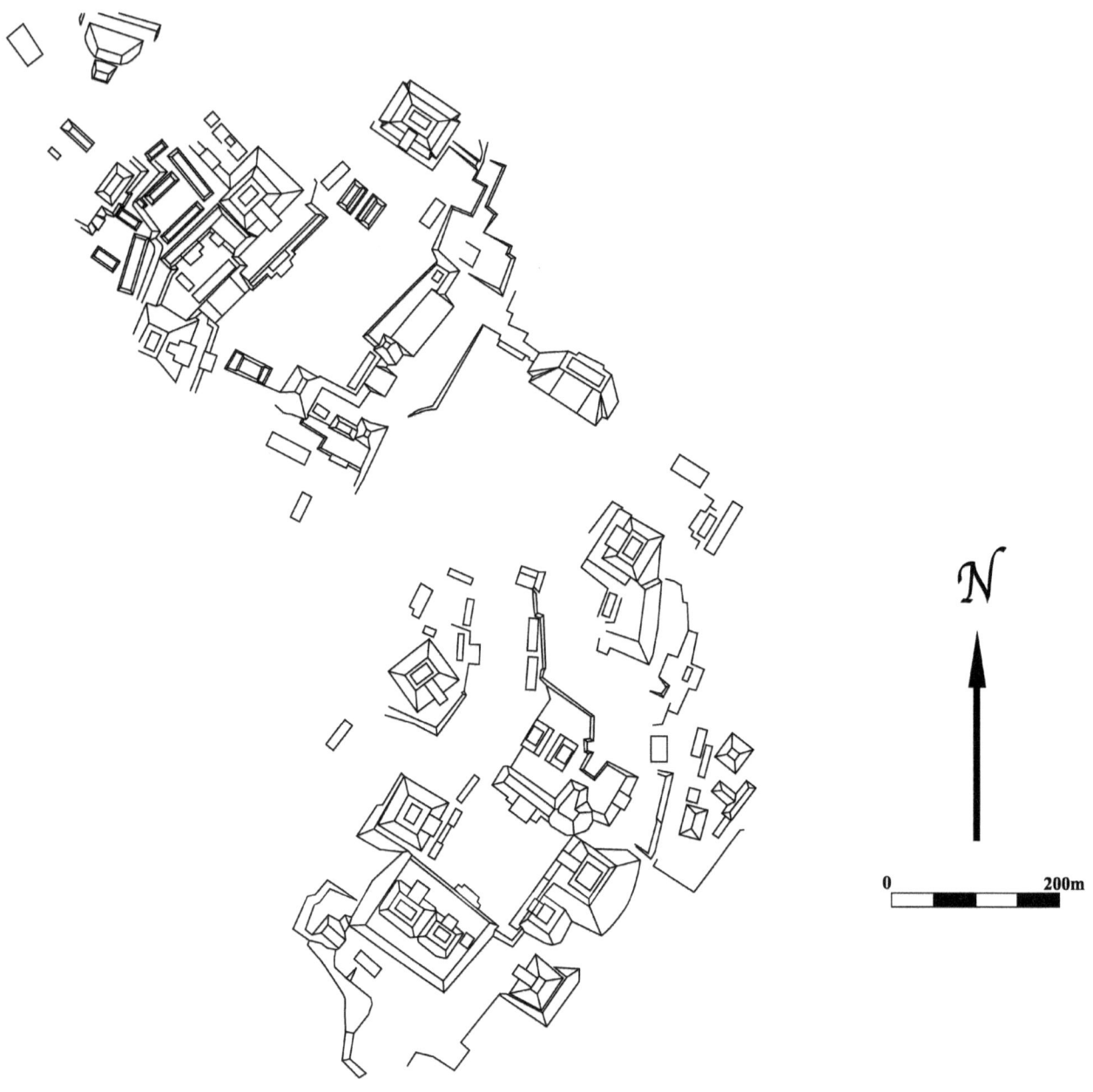

FIGURE 22. Map of Piedras Negras, Guatemala (2000:138).

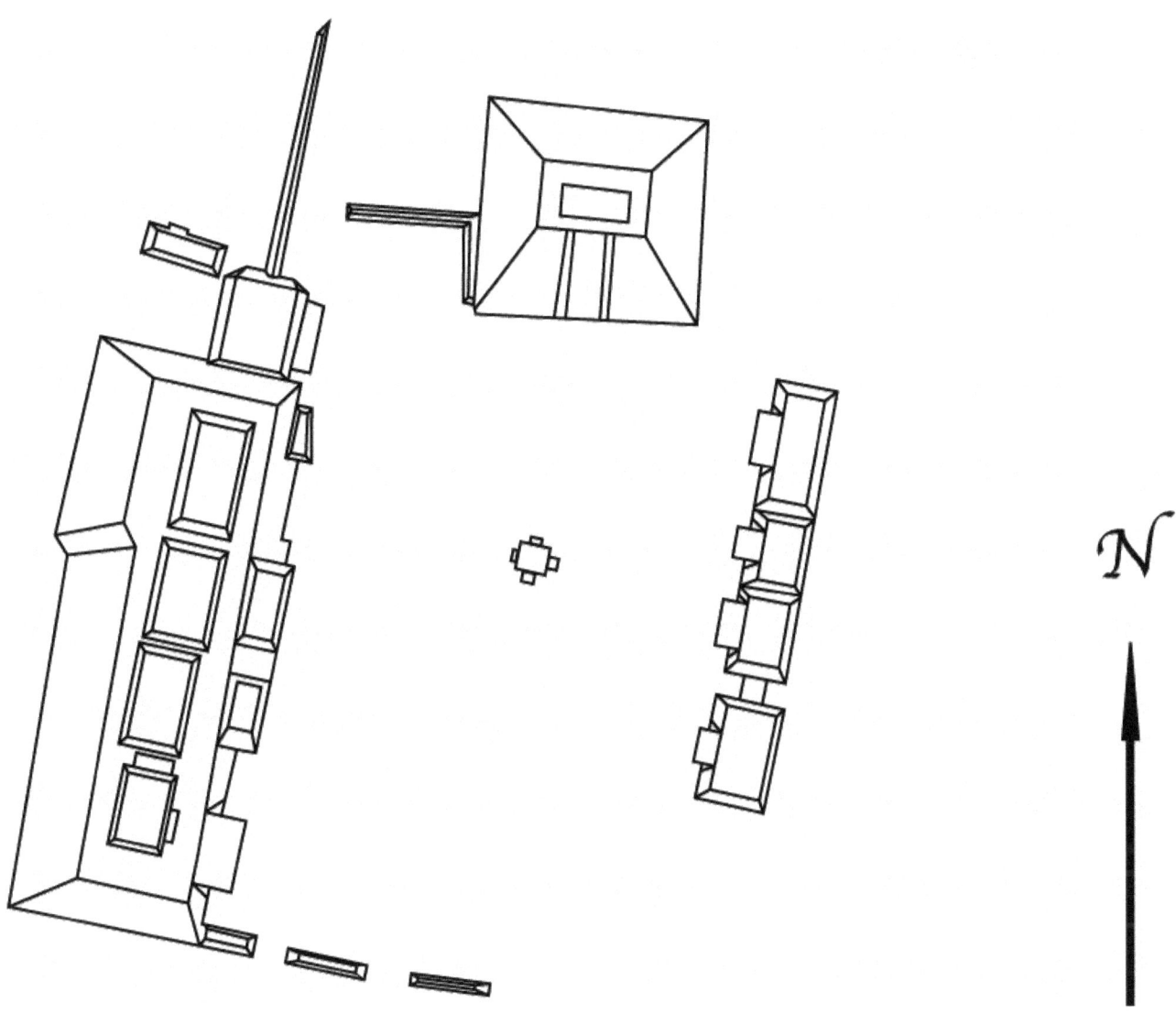

FIGURE 23. Map of Pomona, Mexico (after Tour by Mexico 2003).

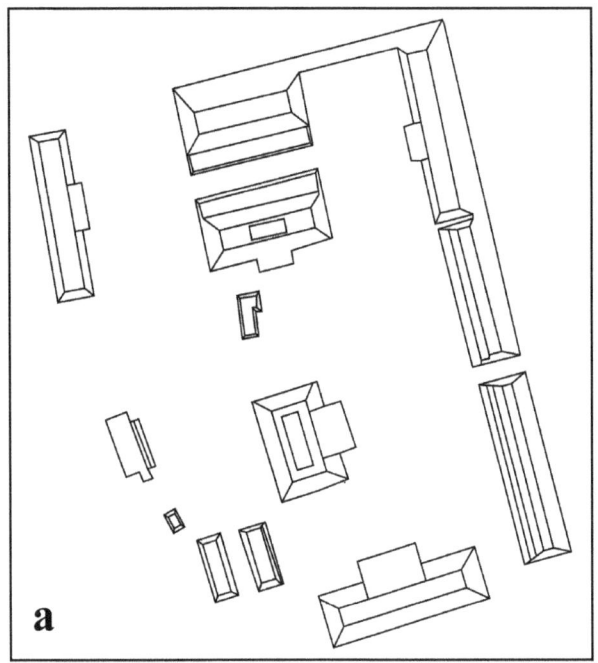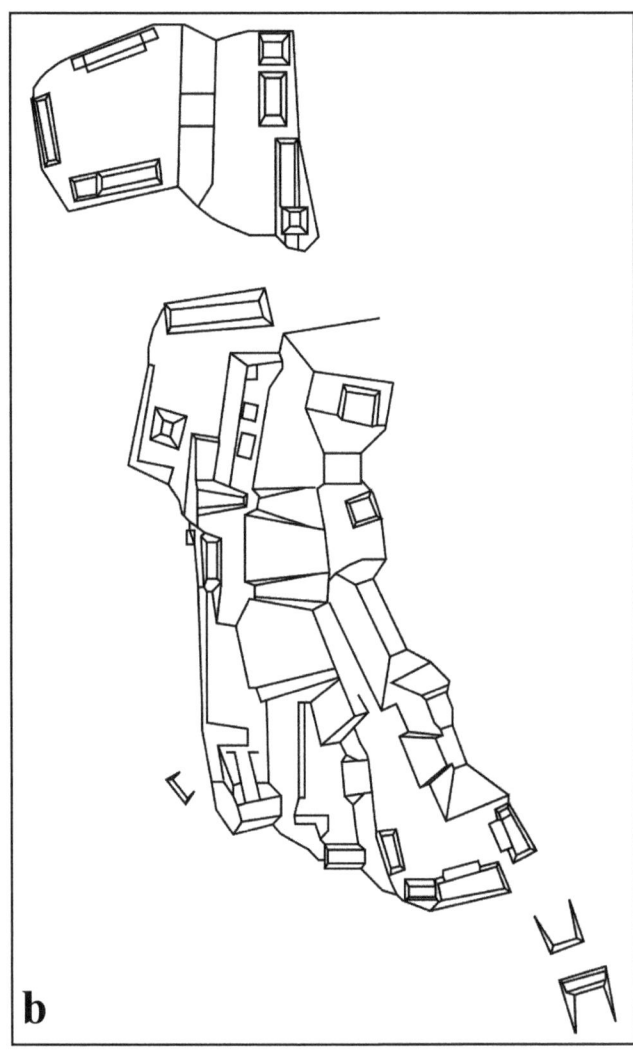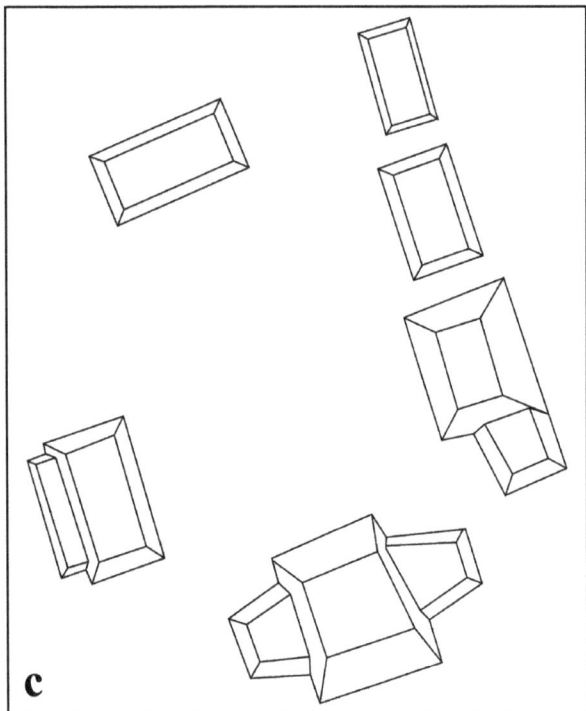

FIGURE 24. Mapped sections of Pusilha, Belize: (a) Moho Group, (b) Gateway Hill Acropolis; (c) Stela Group (after Braswell 2002). Maps not shown to scale.

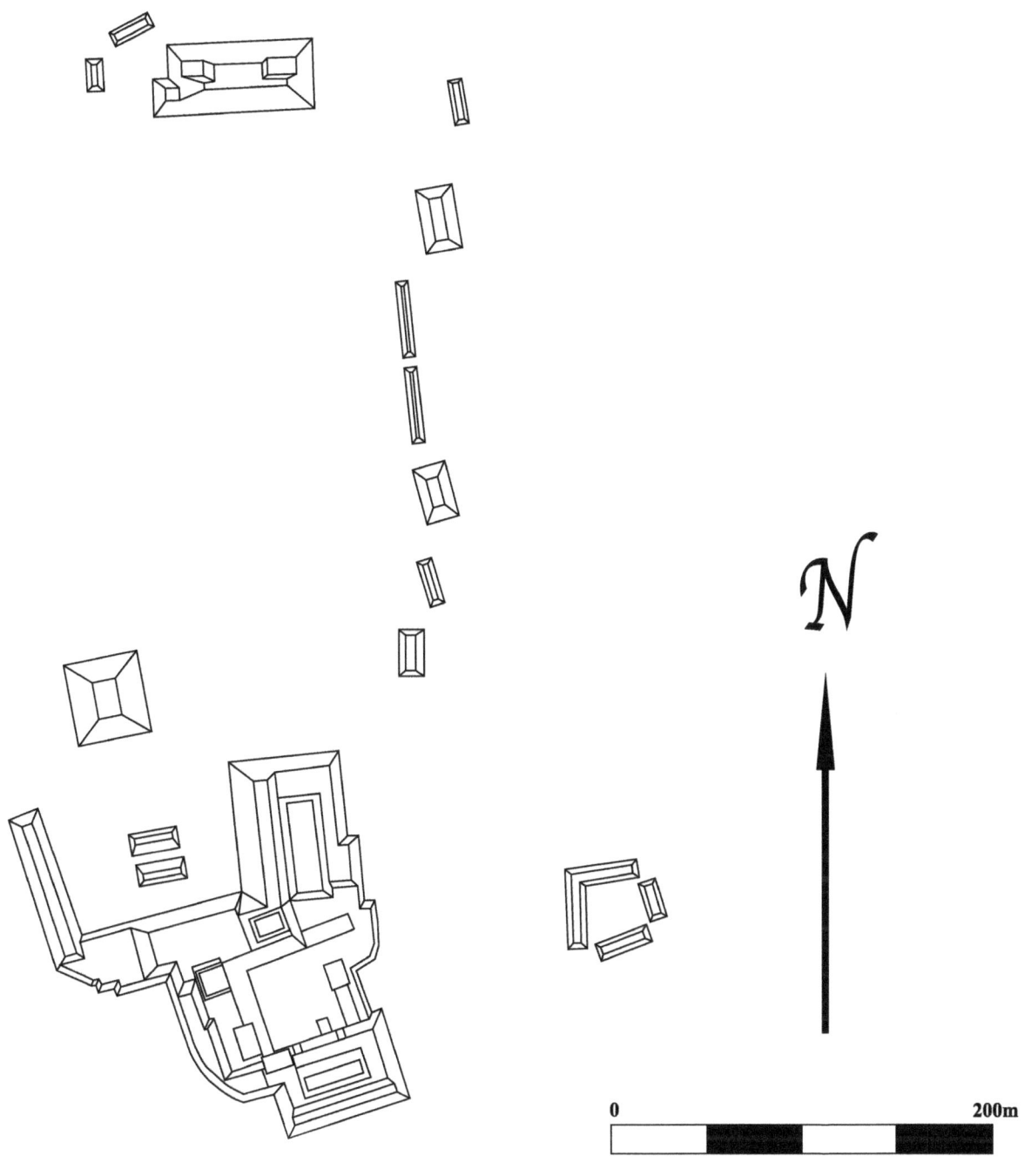

FIGURE 25. Map of Qurigua, Guatemala (after Martin and Grube 2000:214).

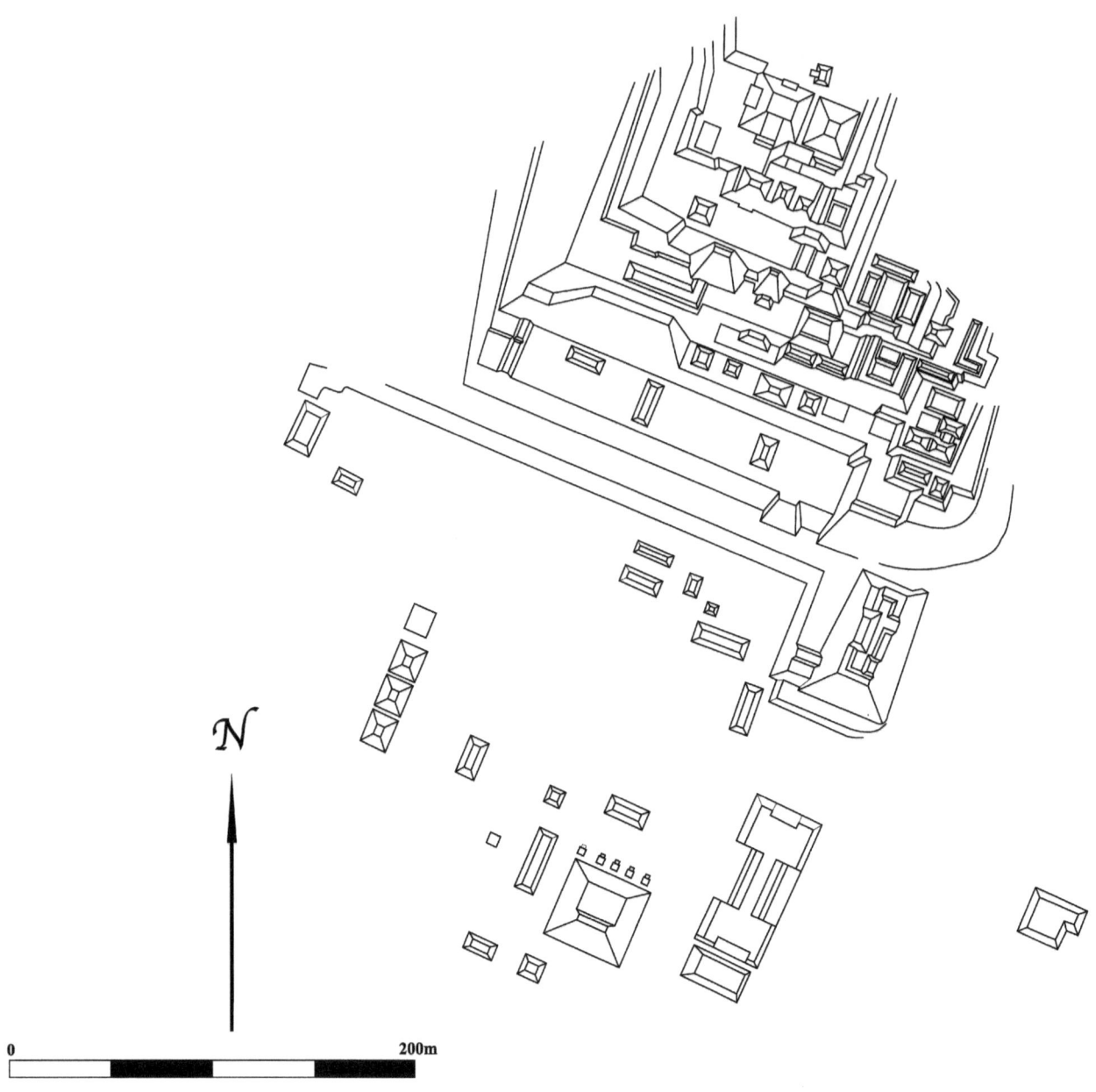

FIGURE 26. Map of Tonina, Mexico (after Martin and Grube 2000:176).

BUILDING IDENTITIES: SOCIO-POLITICAL IMPLICATIONS OF ANCEINT MAYA CIVIC PLANS

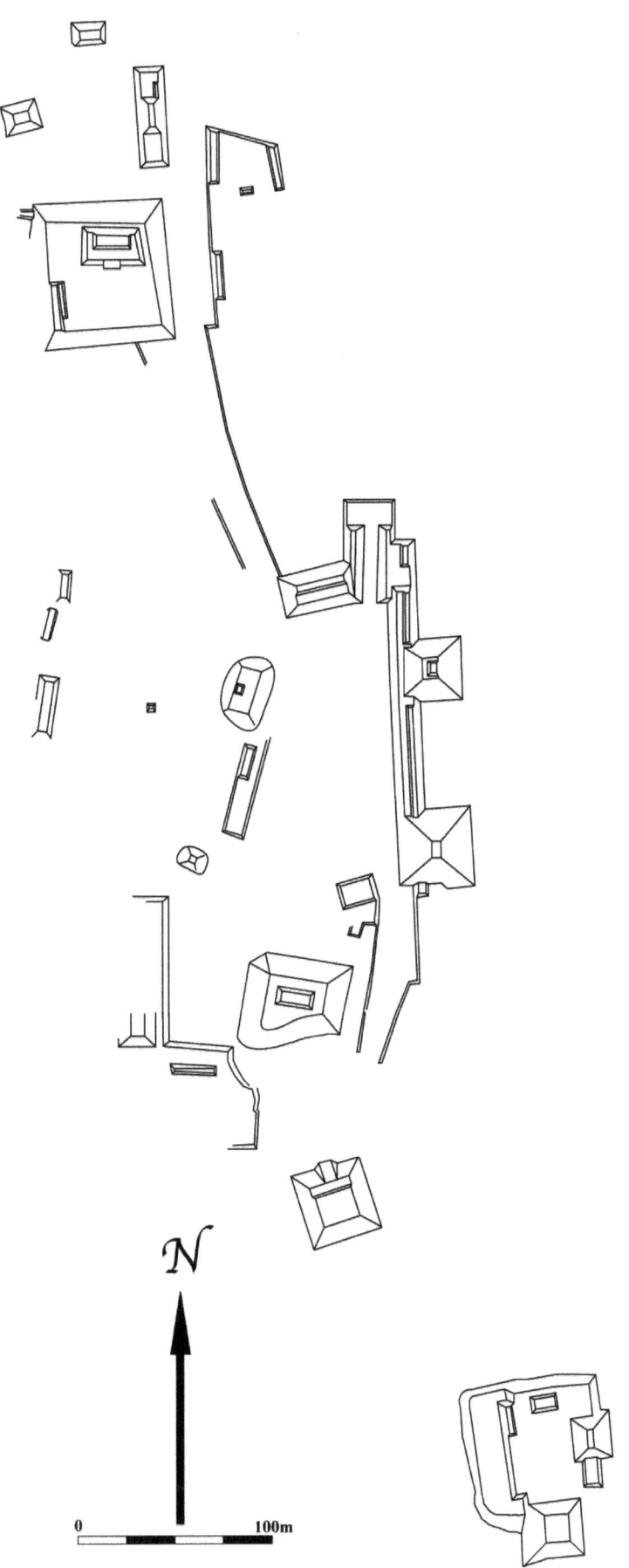

FIGURE 27. Map of Ucanal, Guatemala (after Graham 1980).

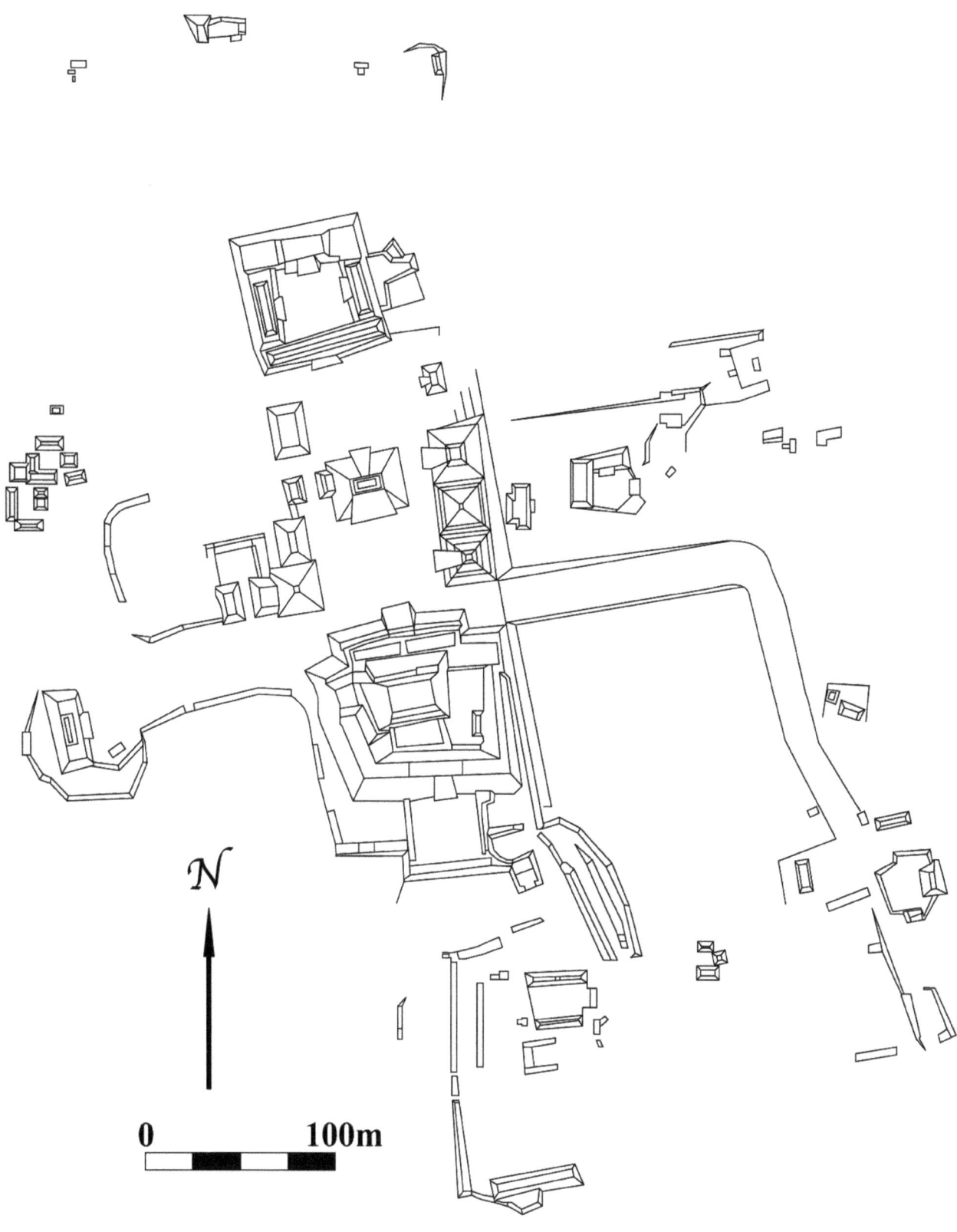

FIGURE 28. Map of Xunantunich, Belize (after LeCount 1999:245).

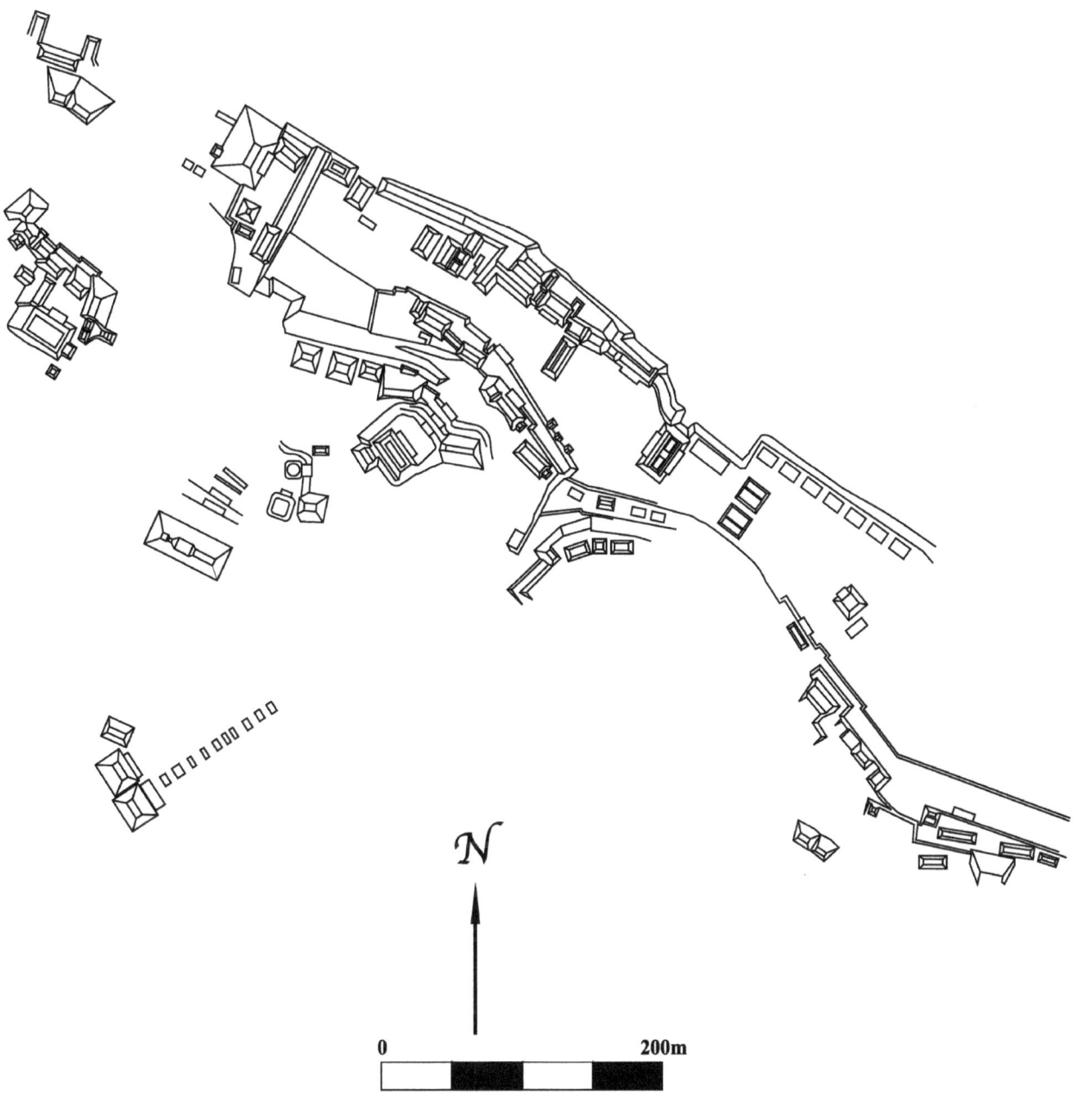

FIGURE 29. Map of Yaxchilan, Mexico (after Martin and Grube 2000:116).

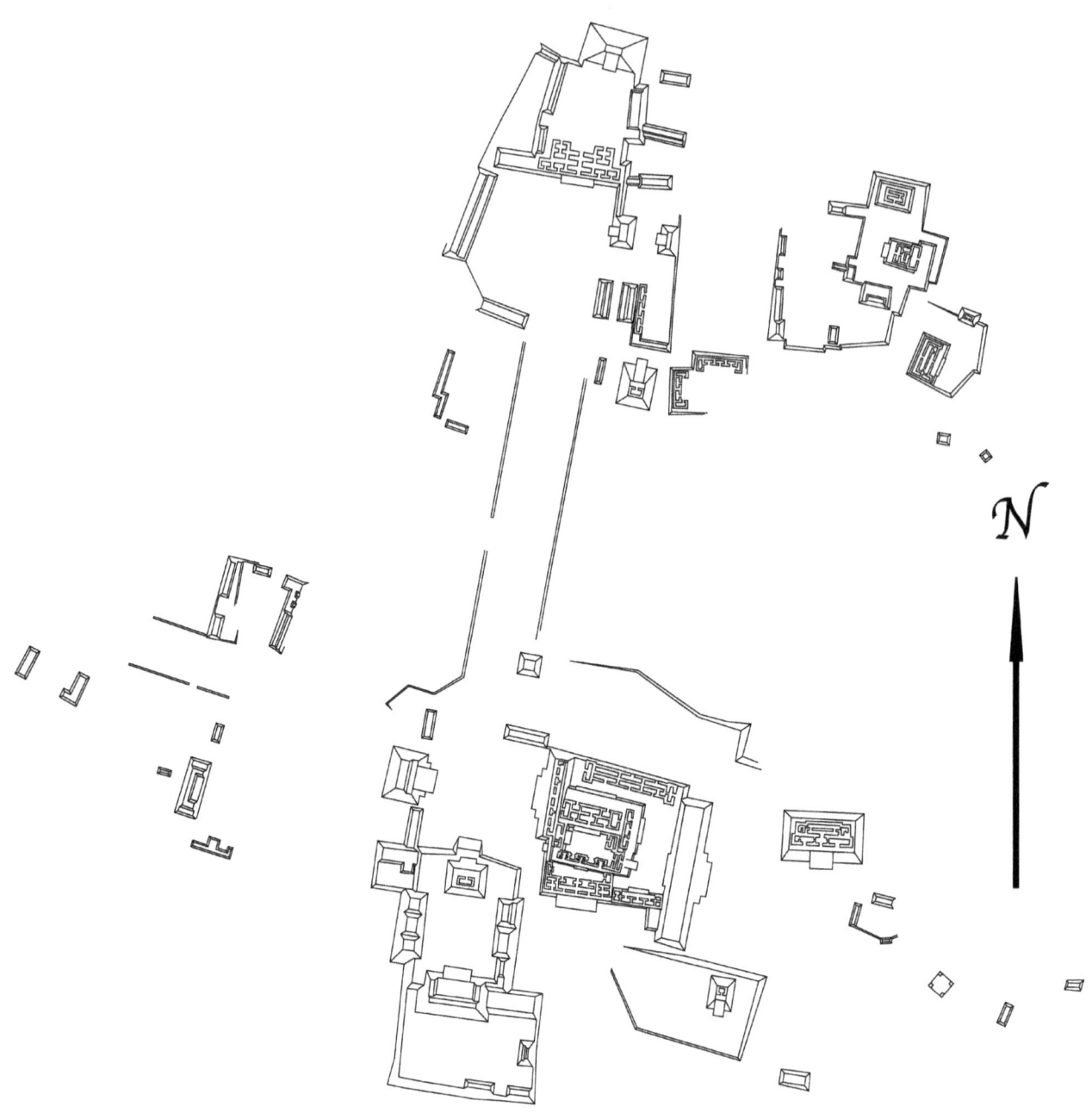

FIGURE 30. Map of Uaxactun, Guatemala (after Von Euw and Graham 1984).

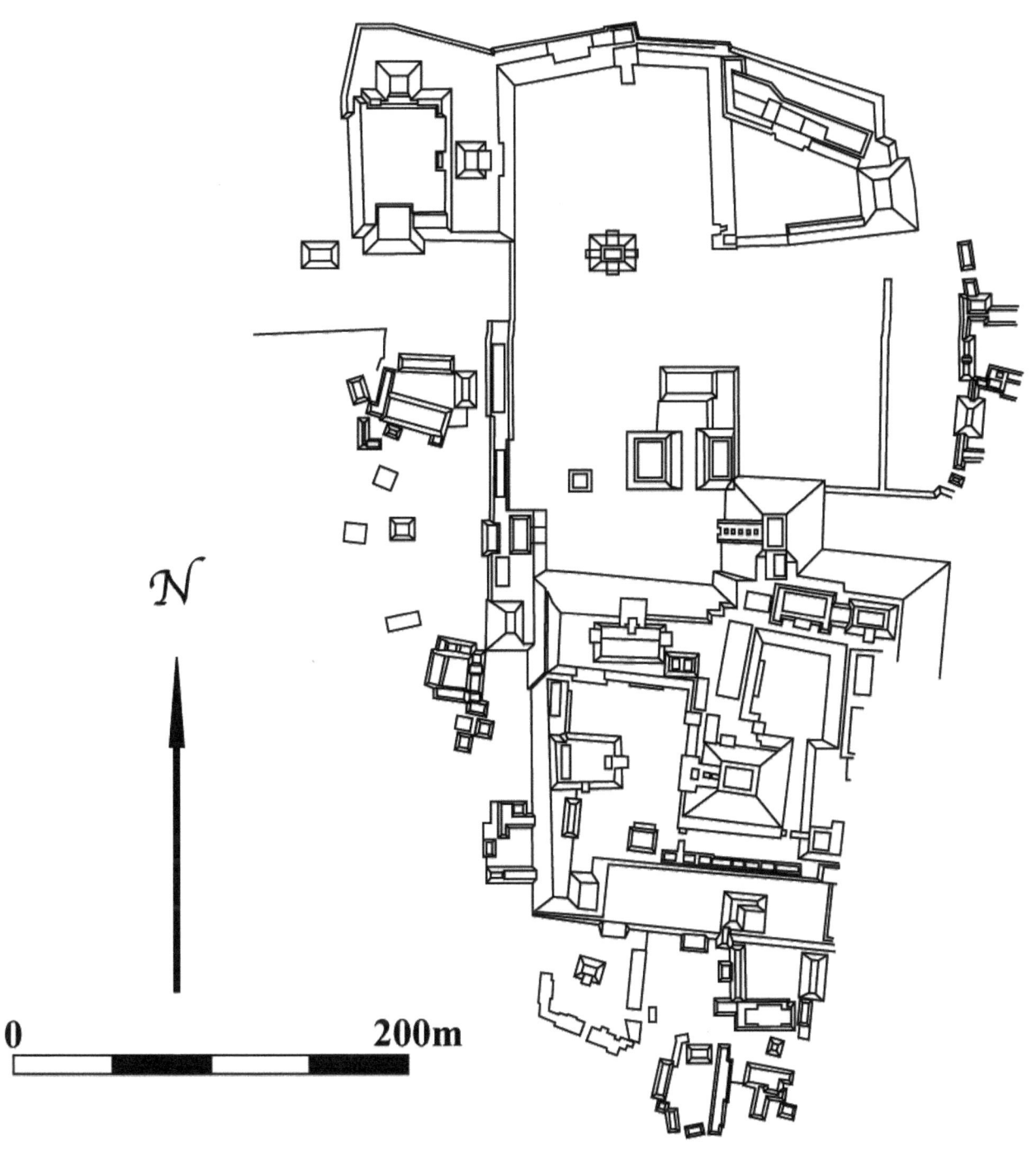

FIGURE 31: Map of Copan, Honduras (after Martin and Grube 2000:190).

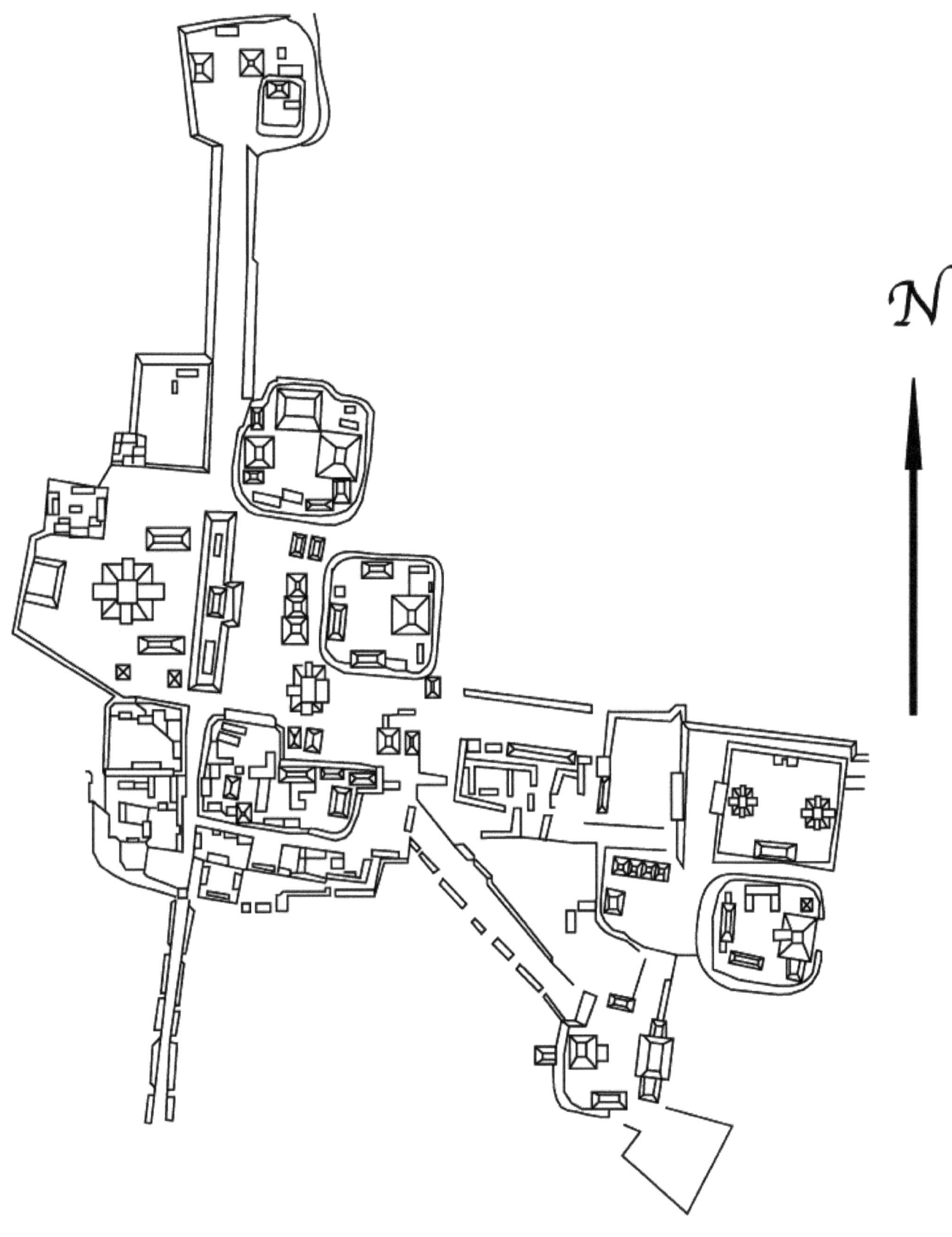

FIGURE 32: Map of Yaxha, Guatemala (after Einführung in das Archäologische Grossprojekt Triangulo Cultural 2003).

APPENDIX 1. MAPS ACQUIRED, INDICATING UNKNOWN CITIES

City	Map	City	Map
Tikal	Yes	Palenque	Yes
Motul de San Jose	No	Maasal	No
Copan	Yes	Ixlu	No
Jimbal	Yes	Uaxactun	Yes
Dos Pilas	Yes	Naranjo	Yes
El Peru	No	Yaxa'	Unknown
Caracol	Yes	Cancuen	Yes
Moral	No	Piedras Negras	Yes
La Corona	No	Quirigua	Yes
Tonina	Yes	Oxpemul	No
Nadzcaan	Yes	La Muñeca	Yes
Calakmul	Yes	Arroyo de Piedra	Yes
Itzan	No	Tamarandito	Yes
Aguateca	Yes	Aguas Calientas	No
El Caribe	No	Chapayal	No
Ahkul	No	El Chorro	No
Yaxchilan	Yes	Yaxha	Yes
Holmul	Yes	Xunantunich	Yes
B'ital	Unknown	Tuub'al	Unkown
Ucanal	Yes	Yootz	Unknown
Sa'kha	Unknown	K'inickil Kab'	Unknown
Dotted Ko	Unknown	Eared Skull	Unknown
Cahal Pichik	Yes	La Rejolla	No
Ixkun	Yes	Ko-Bent Cauc	Unknown
Tzam'	Unknown	Laxtunich	Unknown
La Pasadita	Yes	El Chicozapote	No
Bonampak/ Lacanah	Yes	Hix Witz	Unknown
Maan/Namaan	Unknown	B'uktuun	Unknown
Wak'ab	Unknown	Sanab' Huk'ay	Unknown
Lakamtuun	Unknown	Sak T'zi'	Unknown
Kina'	Unknown	El Cayo	Yes
La Mar	Unknown	Wa-Bird	Unknown
Pomona	Yes	Toktan	Unknown
Oxte'k'uh	Unknown	Anayte	Unknown
Comalcalco	Yes	Pipa'	Unknown
Annak'	Unknown	Pomoy	Unknown
Pusilha	Yes	Nim Li Punit	Yes
Santa Rita	No	Xkuy	Unknown
Annay Te'	No	La Amelia	No

APPENDIX 2. SYNTAGMATIC AND PARADIGMATIC ELEMENTS PRESENT.

City	Tikal Paradigmatic/ Tikal Syntagmatic/ Calakmul Paradigmatic/ Calakmul Syntagmatic
Tikal	NA/ NA/NA/NA
Palenque	1, 4/ 1, 2, 3/ 2/ NA
Jimbal	1/ 1, 3/ 2/ 3
Uaxactun	1/ 1, 2, 3, 4, 5, 7/ 2/ 3
Dos Pilas	NA/ 1, 4/ NA/ 1
Naranjo	1/ 1, 7/ 1, 2/ 2, 3, 4
Caracol	1/ 1, 2, 4, 7/ 2/ 1, 3, 4
Cancuen	NA/ NA/NA/NA
Piedras Negras	1/ 2, 3/ 2/ 1, 4
Quirigua	1/ 1, 4/ 2/ 1
Tonina	NA/ 1/ NA/ 1
Nadzcaan	1/ 1, 4/ 1, 2/ 2, 3, 4
La Muñeca	1/ 5/ 2/ 1, 3, 4
Calakmul	NA/ NA/NA/NA
Arroyo de Piedra	NA/ NA/NA/1, 4
Tamarandito	1/ 4/ 2/ 1, 3, 4
Aguateca	NA/ NA/NA/NA
Yaxchilan	NA/ 1, 2/ NA/ 2
Holmul	NA/ 1, 5/ NA/ 3, 4
Xunantunich	1, 3/ 1, 7/ 1, 2/ 1, 2, 3, 4
Yaxha	1, 2/ 1, 3, 4, 5, 6, 7/ 2/ 2, 3
Ucanal	1/ 1, 4/ 2/ 1, 2, 4
Cahal Pichik	NA/ 1/ NA/ 2, 4
Ixkun	1/ 1, 3, 4, 7/ 2/ 2, 3
La Pasadita	NA/ NA/NA/NA
Bonampak/ Lacanah	NA/ NA/NA/NA
El Cayo	1/ NA/ 2/ NA
Pomona	1, 4/ 2/ 2/ 2
Comalcalco	NA/ 1, 2/ NA/ NA
Pusilha	NA/ 4/ NA/ NA
Nim Li Punit	1/ 1/ 2/ 2

BIBLIOGRAPHY

Andrews, George
1975 *Maya Cities: Placemaking and Urbanization.* University of Oklahoma Press, Normal.

2001 Comalcalco. In *Archaeology of Ancient Mexico and Central America: An Encyclopedia*, edited by Susan Toby Evans and David Webster, pp. 166-168. Garland, New York.

Adams, Richard E. W., and Richard C. Jones
1981 Spatial Patterns and Regional Growth among Classic Maya Cities. *American Antiquity* 46:301-322.

Aimers, James J.
1993 *Messages from the Gods: An Hermenutic Analysis of the Maya E-Group Complex.* Unpublished M.A. thesis, Department of Anthropology, Trent University, Peterborough.

Ashmore, Wendy
1986 Peten Cosmology in the Maya Southeast: An Analysis of Architecture and Settlement Patterns at Classic Quirigua. In *The Southeast Maya Periphery*, edited by Patricia A. Urban and Edward M. Schortman, pp. 35-49. University of Texas Press, Austin.

1989 Construction and Cosmology: Politics and Ideology in Lowland Maya Settlement Patterns. In *Word and Image in Maya Culture: Explorations in Language, Writing, and Representation*, edited by William F. Hanks and Don S. Rice, pp. 272-286. University of Utah Press, Salt Lake City.

1991 Site-Planning Principles and Concepts of Directionality among the Ancient Maya. *Latin American Antiquity* 2:199-226.

1992 Deciphering Maya Architectural Plans. In *New Theories on the Ancient Maya*, edited by E.C. Darien and Robert J. Sharer, pp. 173-184. The University Museum, University of Pennsylvania, Philadelphia.

2005 The Idea of a Maya Town. In *Structure and Meaning in Human Settlement*, edited by Tony Atkin and Joseph Rykwert, pp. 35-54. University of Pennsylvania Museum Publications, Philadelphia.

Ashmore, Wendy, and A. Bernard Knapp (editors)
1999 *Archaeologies of Landscape: Contemporary Perspectives.* Blackwell, Oxford.

Ashmore, Wendy and Jeremy A. Sabloff
2002 Spatial Orders in Maya Civic Plans. *Latin American Antiquity* 13:201-215.

2003 Interpreting Ancient Maya City Plans: Reply to Smith. *Latin American Antiquity* 14:229-238.

Aveni, Anthony, and H. Hartung
1986 *Maya City Planning and the Calendar.* Transactions of the American Philosophical Society, Vol. 76, Part 7. American Philosophical Society, Philadelphia.

Baines, John, and Norman Yoffee
1998 Order, Legitimacy, and Wealth in Ancient Egypt and Mesopotamia. In *Archaic States,* edited by Gary M. Feinman and Joyce Marcus, pp. 199-260. School of American Research, Santa Fe.

2000 Order, legitimacy, and wealth: setting the terms. In *Order, legitimacy, and wealth in ancient states*, edited by Janet Richards and Mary Van Buren, pp. 13-17. Cambridge University Press, Cambridge.

Bagwell, Elizabeth A.
2006 Specialization, Social Complexity and Vernacular Architecture: A cross cultural study of space and construction. In *Space and Spatial Analysis in Archaeology*, edited by Elizabeth C. Robertson, Jeffrey D. Seibert, Deepika C. Fernandez, and Marc U. Zender, pp. 29-35. University of Calgary Press, Calgary.

Ball, Joseph W. and Jennifer T. Taschek
1991 Late Classic Lowland Maya Political Organization and Central-Place Analysis: New Insights from the Upper Belize Valley. *Ancient Mesoamerica* 2:149-165.

2001 The Buenavista-Cahal Pech Royal Court: Multi-Palace Court Mobility and Usage in a Petty Lowland Maya Kingdom. In *Royal Courts of the Ancient Maya, Vol. 2: Data and Case Studies*, edited by Takeshi Inomata and Stephen D. Houston, pp. 65-200. Westview Press, Boulder.

Barrett, John C.
1999a The Living, the Dead and the Ancestors: Neolithic and Early Bronze Age Mortuary Practices. In *Contemporary Archaeology in Theory: A Reader*, edited by Robert Preucel and Ian Hodder, pp. 394-412. Blackwell, Oxford.

1999b The Mythical Landscapes of the British Iron Age. In *Archaeologies of Landscape: Contemporary Perspectives*, edited by Wendy Ashmore and A. Bernard Knapp, pp. 253-265. Blackwell, Oxford.

Bawden, Garth
1995 The Structural Paradox: Moche Culture as Political Ideology. *Latin American Antiquity* 6:255-273.

Becker, Marshall
1999 *Tikal Report 21. Excavations in Residential Areas of Tikal: Groups with Shrines.* The University Museum, Philadelphia.

Bender, Barbara
1992 Theorising Landscapes, and the Prehistoric Landscapes of Stonehenge. *Man* 27: 735-755.

Berdan, Frances F.
1982 *The Aztecs of Central Mexico: An Imperial Society.*

Holt, Rinehart, and Winston, Fort Worth.

Berlin, Heinrich
1958 El glifo "emblema" en las inscripciones mayas. *Journal de la Société des Américanistes* 47:111-119.

Binford, Lewis R.
1982 Archaeology of Place. *Journal of Anthropological Archaeology* 1:5-31.

Blanton, Richard E., and Lane Fargher
2008 *Collective Action in the Formation of Pre-Modern States*. Fundamental Issues in Archaeology. Springer, New York.

Bourdieu, Pierre
1977 *Outline of a Theory of Practice*. Cambridge University Press, Cambridge.

Brady, James E.
1997 Settlement Configuration and Cosmology: The Role of Caves at Dos Pilas. *American Anthropologist* 99:602-618.

Brady, James E., and Wendy Ashmore
1999 Mountains, Caves, Water: Ideational Landscapes of the Ancient Maya. In *Archaeologies of Landscape: Contemporary Perspectives*, edited by Wendy Ashmore and A. Bernard Knapp, pp. 124-145. Blackwell, Oxford.

Braswell, Geoffrey E.
2001 Pusilhá Archaeological Project. Electronic document. http://www.famsi.org/reports/00029/index.html, accessed April 4, 2003.

2003 Introduction: Reinterpreting Early Classic Interaction. In *The Maya and Teotihuacan: Reinterpreting Early Classic Interaction*, edited by Geoffrey E. Braswell, pp. 1-44. University of Texas Press, Austin.

Braswell, Geoffrey E., J.D. Gunn, M. del R. Domminguez C, William J. Folan, L. Fletcher, A. Morales, and M.D. Glascock
2004 Defining the Terminal Classic at Calakmul, Campeche. In *The Terminal Classic in the Maya Lowlands: Collapse, Transition, and Transformation*, edited by Don S. Rice, Prudence M. Rice, and Arthur A. Demarest. University of Colorado Press, Boulder.

Bricker, Victoria
1981 Directional Glyphs in Maya Inscriptions and Codices. *American Antiquity* 48:347-353.

Brotherson, Gordon
1976 Mesoamerican Description of Space II: Signs for Direction. *Ibero-Amerikanisches Archiv* N.F. Jg 2, H. 1:39-62.

Brotherson, Gordon, and Dawn Ades
1975 Mesoamerican Description of Space I: Myths; Stars and Maps; and Architecture. *Ibero-Amerikanisches Archiv* N.F. 1:279-305.

Carl, Peter
2000 City-Image versus Topography of Praxis. *Cambridge Archaeological Journal* 10:328-335.

Carrasco, Davíd
1990 *Religions of Mesoamerica: Cosmovision and Ceremonial Centres*. Religious Traditions of the World. Harper, San Francisco.

Carrasco, Davíd, Johama Brodan, and Eduardo Matos Moctezuma
1988 *The Great Temple of Tenochtitlan: Centre and Periphery in the Aztec World*. University of California Press, Berkley.

Carrasco, Ramón, and Marc Wolf
1996 Nadzca'an: Una antigua ciudad en el suroeste de Campeche, México. *Mexicon* 18:70-86.

Charlton, Thomas A., and Deborah L. Nichols
1997 The City-State COncept: Development and Applications. In *The Archaeology of City-States: Cross-Cultural Approaches*, edited by Deborah L. Nichols and Thomas A. Charlton, pp. 1-14. Smithsonian Institution Press, Washington and London.

Chase, Arlen F.
2004 Polities, Politics, and Social Dynamics: "Contextualizing" the Archaeology of the Belize Valley and Caracol. In *The Ancient Maya of the Belize Valley: Half a Century of Archaeological Research*, edited by James F. Garber pp. 320-334. University Press of Florida, Gainesville.

Chase, Arlen F. and Diane Z. Chase
1996a More than Kin and King: Centralized Political Organization among the Late Classic Maya. *Current Anthropology* 37:803-810.

1996b A Mighty Maya Nation: How Caracol Built an Empire by Cultivating its "Middle Class". *Archaeology* 49:66-72.

1998 Late Classic Maya Political Structure, Polity Size, and Warfare Arenas. In *Anatomia de Una Civilizacion. Aproximaciones Interdisciplinarias a la Cultura Maya*. Publicaciones de la Sociedad Española de Estudios Mayas Num. 4, edited by Andrés Ciudad Ruiz, Yolanda Fernanández Marquínez, José Miguel García Campillo, M. Josefa Iglesias Ponce de León, Alfonso Lacadena García-Gallo, and Luis T. Sanz Castro. Sociedad Española de Estudios Mayas, Madrid.

2001a The Royal Courts of Caracol, Belize: Its Palaces and People. In *Royal Courts of the Ancient Maya, Vol.2: Data and Case Studies*, edited by Takeshi Inomata and Stephen D. Houston, pp. 102-137. Westview Press, Boulder.

2001b Ancient Maya Causeways and Site Organization at Caracol, Belize. *Ancient Mesoamerica* 12:273-281.

Chase, Diane Z., and Arlen F. Chase
1998 The Architectural Context of Caches, Burials, and Other Ritual Activities for the Classic Period Maya (as Reflected at Caracol, Belize). In *Function and Meaning in Classic Maya Architecture*, edited by Stephen D. Houston, pp. 239-332. Dumbarton Oaks, Washington.

Chase, Diane Z., and Arlen F. Chase (editors)
1987 *Investigations at the Classic Maya City of Caracol, Belize: 1985-1987*. Pre-Columbian Art Research Institute Monograph 1. Pre-Columbian Art Research Institute, San Francisco.

1994 *Studies in the Archaeology of Caracol, Belize*. Pre-Columbian Art Research Institute Monograph 7. Pre-Columbian Art Research Institute, San Francisco.

Chase, Diane Z., Arlen F. Chase, and William A. Haviland
1990 The Classic Maya City: Reconsidering the "Mesoamerican Urban Tradition". *American Anthropologist* 92:499-506.

Childe, Vere Gordon
1939 *The Dawn of European Civilisation*. Kegan Paul, London.

Chrisomalis, Stephen
2006 Comparative Archaeology: An Unheralded Cross-Cultural Method. In *The Archaeology of Bruce Trigger: Theoretical Empiricism*, edited by Ronald F. Williamson and Michael S. Bisson, pp. 36-51. McGill-Queen's University Press, Montreal and Kingston.

Christie, Jessica Joyce (editor)
2003 *Maya Palaces and Elite Residences: An Interdisciplinary Approach*. University of Texas Press, Austin.

Clarke, David L.
1968 *Analytical Archaeology*. Methuen and Co., Ltd., London.

Coe, Michael D.
1992 *Breaking the Maya Code*. Thames and Hudson, New York.

Coggins, Clemency Chase
1980 The Shape of Time: Some Political Implications of a Four-Part Figure. *American Antiquity* 45:727-739.

Cook, Duncan E., Brigitte Kovacevich, Timothy Beach, and Ronald Bishop
2006 Deciphering the inorganic chemical record of ancient human activity using ICP-MS: a recinnaissance study of late Classic soil floors at Cancuén, Guatemala. *Journal of Archaeological Science* 33:628-640.

Cooney, Gabriel
1999 Social Landscapes in Irish Prehistory. In *The Archaeology and Anthropology of Landscape: Shaping your Landscape*, edited by Peter J. Ucko and Robert Layton, pp. 46-64. Routledge, London.

Cosgrove, Daniel E (editor)
1988 *The Iconography of Landscape: Essays on the Symbolic Representation, Design,and Use of Past Environments*. Cambridge University Press, Cambridge.

Cowgill, George L.
2000 Intentionality and Meaning in the Layout of Teotihuacan, Mexico. *Cambridge Archaeological Journal* 10:358-361.

2004 Origins and Development of Urbanism: Archaeological Perspectives. *Annual Review of Anthropology* 33:525-549.

Culbert, T. Patrick
1988 Political History and the decipherment of Maya glyphs. *Antiquity* 62:135-152.

1991 Polities in the northeast Petén, Guatemala. In *Classic Maya Political History: Hieroglyphic and Archaeological Evidence*, edited by T. Patrick Culbert, pp. 128-146. Cambridge University Press, Cambridge.

Dahlin, Bruce H., Christopher T. Jensen, Richard E. Terry, David R. Wright, and Timothy Beach
2007 In Search of an Ancient Maya Market. *Latin American Antiquity* 18:363-384.

Demarest, Arthur A.
1992 Ideology in Ancient Maya Cultural Evolution: The Dynamics of Galactic Polities. In *Ideology and Pre-Columbian Civilizations*, edited by Arthur Demarest and G. Conrad, pp.135-158. School of American Research Press, Santa Fe.

2004 *Ancient Maya: The Rise and Fall of a Rainforest Civilization*. Cambridge University Press, Cambridge.

Demarest, Arthur A., Kim Morgan, Claudia Wolley, and Héctor Escobedo
2003 The Political Acquisition of Sacred Geography: The Murciélagos Complex at Dos Pilas. In *Maya Palaces and Elite Residences: An Interdisciplinary Approach*, edited by Jessica Joyce Christie, pp. 120-153. University of Texas Press, Austin.

Driver, W. David and Phil Wanyerka
2002 Creation Symbolism in the Architecture and Ritual at Structure 3, Blue Creek, Belize. *Mexicon* 24:6-8.

Dunning, Nicholas P., Vernon Scarborough, Fred Valdez, Jr., Sheryl Luzzadder-Beach,
Timothy Beach and John G. Jones
1999 Temple Mountains, Sacred Lakes, and Fertile Fields: Ancient Maya Landscapes in Northwestern Belize. *Antiquity* 73:650-60.

Einführung in das Archäologische Grossprojekt Triangulo Cultural
2003 Yaxha. Electronic document. http://www.yaxha.de/WWWyaxhaDEyaxha.htm, accessed December 5, 2003.

Eliade, Mircea
1959 *The Sacred and the Profane: the Nature of Religion*. Harcourt, Brace and World, New York.

Erickson, Paul A., and Liam D. Murphy
1998 *A History of Anthropological Thought*. Broadview Press, Peterborough.

Escobedo, Héctor L.
1997 Arroyo de Piedra: Sociopolitical Dynamics of a Secondary Centre in the Petexbatun Region. *Ancient Mesoamerica* 8:307-320.

Estrada-Belli, Francisco
2000 Archaeological Investigations at Holmul, Guatemala. Electronic document. http://www.famsi.org/reports/980101/images/fig03.jpg, accessed November 12, 2003. Foundation for the Advancement of Mesoamerican Studies, Inc.

Fahsen, Fredrico
2003 Rescuing the Origins of Dos Pilas Dynasty: A Salvage of Hieroglyphic Stairway # 2, Structure L5-49. Electronic document. http://www.famsi.org/reports/01098/index.html, accessed March 12, 2003. Foundation for the Advancement of Mesoamerican Studies, Inc.

Fash, Barbara W.
1992 Late Classic Architectural Themes in Copan. *Ancient Mesoamerica* 3: 89-104.

Fash, William L.
1988 A New Look at Maya Statecraft from Copan, Honduras. *Antiquity* 62:157-159.

1998 Dynastic Architectural Programs: Intention and Design in Classic Maya Buildings at Copan and Other Sites. In *Function and Meaning in Classic Maya Architecture*, edited by Stephen D. Houston, pp. 223-270. Dumbarton Oaks, Washington.

2002 Religion and Human Agency in Ancient Maya History: Tales from the Hieroglyphic Stairway. *Cambridge Archaeological Journal* 12:5-19.

Feinman, Gary M, and Joyce Marcus
1998 Introduction. In *Archaic States*, edited by Joyce Marcus and Gary M. Feinman, pp. 3-14. School of American Research Press, Santa Fe.

Flannery, Kent
1968 Archaeological Systems Theory and Early Mesoamerica. In *Anthropological Archaeology in the Americas*, edited by Betty J. Meggars, pp. 67-87. Anthropological Society of Washington, Washington.

Flannery, Kent, and Joyce Marcus
1999 Cognitive Archaeology. In *Contemporary Archaeology in Theory: A Reader*, edited by Robert Preucel and Ian Hodder, pp. 350-363. Blackwell, Oxford.

Fletcher, R.
1992 Time Perspectivism, Annales and the Potential of Archaeology. In *Archaeology, Annales, and Ethnohistory*, edited by A. Bernard Knapp, pp 35-49. Cambridge University Press, Cambridge.

Folan, William J.
1992 Calakmul, Campeche: a Centralized Urban Administrative Centre in the Northern Petén. *World Archaeology* 24:158-168.

Folan, William J., Joel D. Gunn, and María del Rosario Domínguez Carrasco
2001 Triadic Temples, Central Plazas and Dynastic Palaces: A Diachronic Analysis of the Royal Court Complex, Calakmul, Campeche, Mexico. In *Royal Courts of the Ancient Maya, Vol. 2: Data and Case Studies*, edited by Takeshi Inomata and Stephen D. Houston, pp. 223-265. Westview Press, Boulder.

Folan, William J., Joyce Marcus, Sophia Pincemin, María del Rosario Domínguez Carrasco, Laraine Fletcher and Abel Morales López
1995 Calakmul: New Data from an ancient Maya Capital in Campeche, Mexico. *Latin American Antiquity* 6:310-334.

Folan, William J., Joyce Marcus and W. Frank Miller
1995 Verification of a Maya Settlement Model through Remote Sensing. *Cambridge Archaeology Journal* 5:277-301.

Ford, Anabel
1991 Economic Variation of Ancient Maya Residential Settlement in the Upper Belize River Area. *Ancient Mesoamerica* 2:35-46.

Fox, John W.
1987 *Maya Postclassic State Formation: Segmentary Lineage Migration in Advancing Frontiers*. Cambridge University Press, Cambridge.

1989 On the Rise and Fall of *Tuláns* and Maya Segmentary States. *American Anthropologist* 91:656-681.

Fox, John W., and Garrett W. Cook
1996 Constructing Maya Communities: Ethnography for Archaeology. *Current Anthropology* 37:811-820.

Fox, John W., Garrett W. Cook, Arlen F. Chase and Diane Z. Chase
1996 Segmentary versus Centralized States among the Ancient Maya. *Current Anthropology* 37:795-801.

Fox, Richard
1977 *Urban Anthropology*. Prentice Hall, Englewood

Cliffs.

Freidel, David A.
1983 Political Systems in Lowland Yucatán: Dynamics and Structure in Maya Settlement. In *Prehistoric Settlement Patterns: Essays in Honor of Gordon R. Willey*, edited by Evon Z. Vogt and Richard M. Leventhal, pp. 375-386. University of New Mexico Press and Peabody Museum of Archaeology and Ethnography, Harvard University, Cambridge.

1986 Maya Warfare: An Example of Peer Polity Interaction. In *Peer-polity Interaction and Socio-political Change*, edited by Colin Renfrew and John F. Cherry, pp. 93-108. Cambridge University Press, Cambridge.

1992 The Trees of Life: Ahau as Idea and Artifact in Classic Maya Civilization. In *Ideology and Pre-Columbian Civilisations*, edited by Arthur Demarest and G. Conrad, pp. 115-133. School of American Research, Santa Fe.

2008 Maya Divine Kingship. In *Religion and Power: Divine Kingship in the Ancient World and Beyond*, edited by Nicole Brisch, pp. 191-206. The Oriental Institute of the University of Chicago, Chicago.

Freidel, David A. and Linda Schele
1988a Symbol and Power: A History of the Lowland Maya Cosmogram. In *Maya Iconography*, edited by Elizabeth P. Benson and Gillett G. Griffen, pp. 44-93. Princeton University Press, Princeton

1988b Kingship in the Late Preclassic Maya Lowlands: The Instruments and Places of Ritual Power. *American Anthropologist* 90:547-567.

Freidel, David A., Linda Schele, and Joy Parker
1993 *Maya Cosmos: Three Thousand Years on the Shaman's Path*. William Morrow, New York.

Freidel, David A., and Charles Suhler
1999 The Path of Life: Towards a Functional Analysis of Ancient Maya Architecture. In *Mesoamerican Architecture as a Cultural Symbol*, edited by Jeff K. Kowalski, pp. 250-273. Oxford University Press, New York.

Fritz, John M.
1986 Vijayanagara: Authority and Meaning of a South Indian Imperial Capital. *American Anthropologist* 88:44-55.

Fritz, John M., and George Michell
1987 Interpreting the Plan of a Medieval Hindu Capital, Vijayanagara. *World Archaeology* 19:105-129.

Geertz, Clifford
1980 *Negara: The Theatre State in Nineteenth Century Bali*. Princeton University Press, Princeton.

Giddens, Anthony
1984 *The Constitution of Society: An Outline of the Theory of Structuration*. University of California Press, Berkeley.

Goetz, Delia, and Sylvanus G. Morley
1950 *Popul Vuh: The Sacred Book of the Ancient Quiché Maya*. University of Oklahoma Press, Norman.

Golden, Charles
1999 La Pasadita Archaeological Project. Electronic document. http://www.famsi.org/report/97042 , accessed May 23, 2003. Foundation for the Advancement of Mesoamerican Studies, Inc.

Gosden, Christopher
1999 *Anthropology and Archaeology: A Changing Relationship*. Routledge, London.

Graham, Ian
1980 *Corpus of Maya Hieroglyohioc Inscriptions, Vol.2, 3: Ixkun, Ucanal, Ixtutz, Naranjo*. Peabody Museum of Archaeology and Ethnology, Harvard University, Cambridge.

Grube, Nickolai
2000 The City-States of the Maya. In *A Comparative Study of Thirty City-State Cultures: An Investigation Conducted by the Copenhagen Polis Centre*, edited by Mogens Herman Hansen, pp. 547-565. The Royal Danish Academy of Sciences and Letters, Copenhagen.

Hammond, Norman
1974 The Distribution of Late Classic Maya Major Ceremonial Centres in the Central Area. In *Mesoamerican Archaeology: New Approaches*, edited by Norman Hammond, pp. 313-334. University of Texas Press, Austin.

Hansen, Mogens Herman
2000 Introduction: The Concepts of City-State and City-State Culture. In *A Comparative Study of Thirty City-State Cultures: An Investigation Conducted by the Copenhagen Polis Centre*, edited by Mogens Herman Hansen, pp. 11-34. The Royal Danish Academy of Sciences and Letters, Copenhagen.

Hardy, Joyce
1968 *Urban Planning in Pre-Columbian America*. George Braziller, New York.

Harrison, Peter
2001 Thrones and Throne Structures in the Central Acropolis of Tikal as an Expression of Royal Court. In *Royal Courts of the Ancient Maya, Vol.2: Data and Case Studies*, edited by Takeshi Inomata and Stephen D. Houston, pp. 74-101. Westview Press, Boulder.

Haviland, William A.
1969 A New Population Estimate for Tikal, Guatemala. *American Antiquity* 34:429-433.

1970 Tikal, Guatemala and Mesoamerican Urbanism.

World Archaeology 2:186-198.

1977 Dynastic Genealogies From Tikal, Guatemala: Implications For Descent and Political Organization. *American Antiquity* 42:61-67.

1992 Status and Power in Classic Maya Society: The View from Tikal. *American Anthropologist* 94:937-40.

1997 On The Maya State. *Current Anthropology* 38:443-445.

Healy, Paul F., J.D.H. Lambert, J.T. Arnason, and R.J. Hebda
1983 Caracol, Belize: Evidence of Ancient Maya Agricultural Terraces. *Journal of Field Archaeology* 10:773-796.

Henderson, John S.
1997 *The World of the Ancient Maya*, 2nd ed. Cornell University Press, Ithaca.

Hodder, Ian
1982 *Symbolic and Structural Archaeology*. Cambridge University Press, Cambridge.

1987a *Archaeology as Long-Term History*. Cambridge University Press, Cambridge.

1987b The Contextual Analysis of Symbolic Meanings. In *The Archaeology of Contextual Meanings*, edited by Ian Hodder, pp. 1-10. Cambridge University Press, Cambridge.

1992 *Theory and Practice in Archaeology*. Routledge, London.

1999 *The Archaeological Process: An Introduction*. Blackwell, Oxford.

Houk, Brett A.
2003 The Ties That Bind: Site Planning in the Three Rivers Region. In *Heterarchy, Political Economy, and the Ancient Maya: The Three Rivers Region of the East-Central Yucatán Peninsula*, edited by Vernon Scarborough, Fred Valdez Jr., and Nicholas Dunning, pp. 52-63. University of Arizona Press, Tucson.

Houston, Stephen D.
1998 Finding Function and Meaning in Classic Maya Architecture. In *Function and Meaning in Classic Maya Architecture*, edited by Stephen D. Houston, pp.519-538. Dumbarton Oaks, Washington.

Houston, Stephen D., Héctor Escobedo, Richard Terry, David Webster, George Veni, and Kitty F. Emery
2000 Among the River Kings: Archaeological Research at Piedras Negras, Guatemala, 1999. *Mexicon* 22:8-17.

Iannone, Gyles
1993 Ancient Maya Eccentric Lithics: A Contextual Analysis. Unpublished M.A. Thesis, Department of Anthropology, Trent University, Peterborough.

2001 Fall of the House of Minanhá: A Case of Late Classic Political Disruption in West Central Belize. In *Archaeological Investigations in the North Vaca Plateau, Belize: Progress Report of the Third (2001) Field Season*, edited by Gyles Iannone, Ryan Primrose, Adam Menzies and Lisa McPartland, pp. 127-133. Social Archaeology Research Program, Department of Anthropology, Trent University, Peterborough.

2002 Annales History and the Ancient Maya State: Some Observations on the "Dynamic Model". *American Anthropologist* 104:68-78.

Ingold, Tim
1993 The Temporality of the Landscape. *World Archaeology* 25:152-174.

Inomata, Takeshi
2001 King's People: Classic Maya Courtiers in a Comparative Perspective. In *Royal Courts of the Ancient Maya, Volume 1: Theory, Comparison, and Synthesis*, edited by Takeshi Inomata and Stephen D. Houston, pp. 27-53.

Inomata, Takeshi and Stephen D. Houston (editors)
2001a *Royal Courts of the Ancient Maya Volume 1: Theory, Comparison, and Synthesis*. Westview Press, Boulder.

2001b *Royal Courts of the Ancient Maya Volume 2: Data and Case Studies*. Westview Press, Boulder.

Isbell, William
2000 What We Should Be Studying: The "Imagined Community" and the "Natural Community". In *The Archaeology of Communities: A New World Perspective*, edited by Marcello A. Canuto and Jason Yaeger, pp. 243-266. Routledge, London.

Johnson, Matthew
1999 *Archaeological Theory: An Introduction*. Blackwell, Oxford.

Jones, Christopher
1982[1969] *The Twin-Pyramid Group Pattern: A Classic Maya Architectural Assemblage at Tikal, Guatemala*. University Microfilms International, Ann Arbor.

Joyce, Arthur A.
2000 The Founding of Monte Albán: Sacred Propositions and Social Practices. In *Agency in Archaeology*, edited by Marcia Ann-Dobres and John Robb, pp. 71-91. Routledge, London.

Kowalski, Jeff.K. (editor)
1999 *Mesoamerican Architecture as a Cultural Symbol*. Oxford University Press, New York.

Knapp, A. Bernard, and Wendy Ashmore.

1999 Archaeological Landscapes: Constructed, Conceptualized, Ideational. In *Archaeologies of Landscape: Contemporary Perspectives*, edited by Wendy Ashmore and A. Bernard Knapp, pp. 1-30. Routledge, Oxford.

Lampl, Paul
1968 *Cities and Planning in the Ancient Near East*. Studio Vista, New York.

Laporte, Juan Pedro
1993 Architecture and Social Change in Late Classic Maya Society: The Evidence from Mundo Perdido, Tikal. In *Lowland Maya Civilisation In The Eighth Century A.D.*, edited by Jeremy A. Sabloff and J.S. Henderson, pp. 299-320. Dumbarton Oaks, Washington.

Leach, Edmund
1976 *Culture and Communication: The Logic by which Symbols are Connected*. Cambridge University Press, Cambridge.

LeCount, Lisa J.
1999 Polychrome Pottery and Political Strategies in Late and Terminal Classic Lowland Maya Society. *Latin American Antiquity* 10:239-258.

2000 Like Water for Chocolate: Feasting and Political Ritual among the Late Classic Maya at Xunantunich, Belize. *American Anthropologist* 103:935-953.

Lee, Thomas A., Jr., and Brian Hayden
1988 *San Pablo Cave and El Cayo on the Usumacinta River, Chipas, Mexico*. Papers of the New World Archaeological Foundation No. 53. New World Archaeological Foundation, Brigham Young University, Provo.

Leventhal, Richard M.
1983 Household Groups and Classic Maya Religion. In *Prehistoric Settlement Patterns: Essays in Honor of Gordon R. Willey*, edited by Evon. Z.Vogt and Richard M. Leventhal, pp. 55-76. University of New Mexico Press and Peabody Museum of Archaeology and Ethnography, Harvard University, Cambridge.

1992 Southern Belize: An Ancient Maya Region. In *Vision and Revision in Maya Studies*, edited by Flora S. Clany and Peter D. Harrison, pp. 125-142. University of New Mexico Press, Albuquerque.

Leventhal, Richard M., and Wendy Ashmore
2004 Xunantunich in a Belize Valley Context. In *The Ancient Maya of the Belize Valley: Half a Century of Archaeological Research*, edited by James F. Garber, pp. 168-179. University Press of Florida, Gainesville.

Lévi-Strauss, Claude
1962 *The Savage Mind*. University of Chicago Press, Chicago.

1967 The Story of Asdiwal. In *The Structural Study of Myth and Totemism*, edited by Edmund R. Leach, translated by N. Mann, pp. 1-48. Tavistock, London.

Looper, Matthew G.
1995 The Three Stones of Maya Creation Mythology at Quiriguá. *Mexicon* 17:24-30.

1999 New Perspectives on the Late Classic Political History of Quiriguá, Guatemala. *Ancient Mesoamerica* 10:263-280.

Loten, H. Stanley, and David Pendergast
1984 *A Lexicon for Maya Architecture*. Royal Ontario Museum, Toronto.

Lucero, Lisa
2003 The Politics of Ritual: The Emergence of Classic Maya Rulers. *Current Anthropology* 44:523-558.

2007 Classic Maya Temples, Politics, and the Voice of the People. *Latin American Antiquity* 18:407-427.

Macri, Martha J.
1994 The Five-door Temples at Piedras Negras and Palenque. *Mexicon* 16:100-102.

Marcus, Joyce
1973 Territorial Organization of the Lowland Classic Maya. *Science* 4089:911-916.

1976 *Emblem and State in the Classic Maya Lowlands*. Dumbarton Oaks, Washington.

1987 *The Inscriptions of Calakmul: Royal Marriage at a Maya City in Campeche, Mexico*. University of Michigan, Museum of Anthropology, Ann Arbor.

1983 On the Nature of the Mesoamerican City. In *Prehistoric Settlement Patterns: Essays in Honor of Gordon R. Willey*, edited by Evon Z. Vogt and Richard M. Leventhal, pp. 195-242. University of New Mexico Press and Peabody Museum of Archaeology and Ethnography, Harvard University, Cambridge.

1992 *Mesoamerican Writing Systems: Propaganda, Myth, and History in Four Ancient Civilisations*. Princeton University Press, Princeton.

1993 Ancient Maya Political Organization. *In Lowland Maya Civilisation In The Eighth Century A.D.*, edited by Jeremy A. Sabloff and John S. Henderson, pp. 111-184. Dumbarton Oaks, Washington.

1998 The Peaks and Valleys of Ancient States: An Extension of the Dynamic Model. In *Archaic States*, edited by Joyce Marcus and Gary M. Feinman, pp. 59-94. School of American Research, Santa Fe.

2009 The Archaeological Evidence for Social Evolution. *Annual Review of Anthropology* 37:251-266.

Marcus, Joyce, and Jeremy A. Sabloff (editors)

2008 *The Ancient City: New Perspectives on Urbanism in the Old and New Worlds*. School for Advanced Research, Santa Fe.

Martin, Simon
2001 Court and Realm: Architectural Signatures in the Classic Maya Southern Lowlands. In *Royal Courts of the Ancient Maya, Vol.1: Theory, Comparison, and Synthesis*, edited by Takeshi Inomata and Stephen D. Houston, pp. 168-194. Westview Press, Boulder.

Martin, Simon and Nickolai Grube
1994 Macro-Political Organisation Amongst Classic Maya Lowland States. Manuscript. London/Bonn.

1995 Maya Superstates. *Archaeology* 48: 41-46

2000 *Chronicle of the Maya Kings and Queens: Deciphering the Dynasties of the Ancient Maya*. Thames and Hudson, London.

Mathews, Peter
1991 Classic Maya Emblem Glyphs. In *Classic Maya Political History: Hieroglyphic and Archaeological Evidence*, edited by T. Patrick Culbert, pp. 19-29. Cambridge University Press, Cambridge.

McAnany, Patricia A.
1995 *Living with the Ancestors: Kinship and Kingship in Ancient Maya Society*. University of Texas Press, Austin.

1998 Ancestors and the Classic Maya Built Environment. In *Function and Meaning in Classic Maya Architecture*, edited by Stephen D. Houston, pp. 271-298. Dumbarton Oaks, Washington.

McAnnany, Patricia A., and Shannon Plank
2001 Perspectives on Actors, Gender Roles, and Architecture at Classic Maya Courts and Households. In *Royal Courts of the Ancient Maya, Vol.1: Theory, Comparison, and Synthesis*, edited by Takeshi Inomata and Stephen D. Houston, pp. 84-129. Westview Press, Boulder.

Miller, Mary Ellen
1998 A Design for Meaning in Maya Architecture. In *Function and Meaning in Classic Maya Architecture*, edited by Stephen D. Houston, pp. 187-222. Dumbarton Oaks, Washington.

Morley, Sylvanus G.
1916[1977] The Supplementary Series in the Maya Inscriptions. In *Anthropological Essays Presented to William Henry Holmes in Honour of his Seventieth Birthday, December 1, 1916*, pp 366-396. AMS Press, New York, for the Peabody Museum of Archaeology and Ethnology, Harvard University, Cambridge.

1935 *Guide Book to the Ruins of Quirigua*. Carnegie Institution of Washington, Washington.

Mosher, Matthew S.
2004 The Minanha' Site Plan and Inter-regional Interaction. Paper presented at 69th Annual Meeting of the Society for American Archaeology, Montréal.

Mosher, Matthew S., and Stefan Jones
2008 Architecture and Authority: Late Classic Civic Layouts and Political Relations in the North Vaca Plateau. Paper presented at the 73rd Annual Meeting of the Society for American Archaeology.

Mounin, Georges
1980 The Semilology of Orientation in Urban Space. *Current Anthropology* 21:491-501.

Nicholson, Irene
1967 *Mexican and Central American Mythology*. Paul Hamlyn, London.

Orr, Heather S.
2001 Procession Rituals and Shrine Sites: The Politics of Sacred Space in the Late Formative Valley of Oaxaca. In *Landscape and Power in Ancient Mesoamerica*, edited by Rex Koontz, Annabeth Headrick, and Kathryn Reese-Taylor, pp. 55-79. Westview Press, Boulder.

Pearson, Mike Parker and Colin Richards
1994 Ordering the World: Perceptions of Architecture, Space, and Time. In *Architecture and Order: Approaches to Social Space*, edited by Mike Parker Pearson and Colin Richards, pp. 1-37. Routledge, London.

Peuramaki-Brown, Meaghan
2003 Next to Every Great Man is One, or More, Great Women: Maya Women and Political Power in the Classic Period. Paper presented at University of Calgary Gender Symposium, Calgary.

Pincemin, Sophia, Joyce Marcus, Lynda Florey Folan, William J. Folan, María del Rosario Domínguez Carrasco and Abel Morales López
1998 Extending the Calakmul Dynasty Back in Time: A New Stela from a Maya Capital in Campeche, Mexico. *Latin American Antiquity* 9:310-327.

Pollock, Adam
2003 Summary of Ancient Maya Terrace Investigations in the Southern Periphery of Minanha'. In *Archaeological Investigations in the North Vaca Plateau, Belize: Progress Report of the Fifth (2002) Season*, edited by Gyles Iannone and James E. Herbert, pp. 85-101. Social Archaeology Research Program, Department of Anthropology, Trent University, Peterborough.

Preziosi, Donald
1979 *The Semiotics of the Built Environment: An Introduction to Architectonic Analysis*. Indiana University Press, Bloomington.

Primrose, J. Ryan
2003 *The Ancient Maya Water Management System at Minanha', West Central Belize*. Unpublished M.A.

Thesis, Department of Anthropology, Trent University, Peterborough.

2004 Social and Physical Considerations of the Water Management System at Minanhá, West Central Belize. Paper presented at the 69th Annual Meeting of the Society for American Archaeology, Montréal.

Proskouriakoff, Tatiana
1960 Historical Implications of a Pattern of Dates at Piedras Negras, Guatemala. *American Antiquity* 25:454-475.

Puleston, Dennis G.
1983 *Settlement Survey of Tikal. Tikal Report No. 13.* The University Museum, Philadelphia.

Pyburn, Anne K.
2008 Pomp and Circumstance before Belize: Ancient Maya Commerce and the New River Conurbation. In *The Ancient City: New Perspectives on Urbanism in the Old and New Worlds*, edited by Joyce Marcus and Jeremy A. Sabloff, pp. 247-272. School for Advanced Research, Santa Fe.

Reents-Budet, Dorie
2003 The Power of Feasting: Classic Maya Economics. Trent University, Archaeological Research Centre Joan Vastokas Distinguished Lecture in Art and Archaeology, Peterborough.

Reese-Taylor, Katherine
2002 Ritual Circuits as Key Elements in Maya Civic Center Design. In *Heart of Creation: The Mesoamerican World and the Legacy of Linda Schele*, edited by Andrea Stone, pp. 143-165. University of Alabama Press, Tuscaloosa.

Reese-Taylor, Katherine, and Rex Koontz
2001 The Cultural Poetics of Power and Space in Ancient Mesoamerica. In *Landscape and Power in Ancient Mesoamerica*, edited by Rex Koontz, Annabeth Headrick, and Kathryn Reese-Taylor, pp. 1-27. Westview Press, Boulder.

Renfrew, Colin, and John F. Cherry (editors)
1986 *Peer-polity Interaction and Socio-political Change.* Cambridge University Press, Cambridge.

Ruppert, Karl, and John Denison, Jr.
1943 *Archaeological Reconnaissance in Campeche, Quintana Roo, and Peten.* Publication 543. Carnegie Institution of Washington, Washington.

Ruppert, Karl, J. Eric S. Thompson, and Tatania Proskouriakoff
1955 *Bonampak, Chiapas, Mexico.* Carnegie Institution of Washington, Washington.

Rykwert, Joseph
1988 [1976] *The Idea of a Town: The Anthropology of Urban Form in Rome, Italy and the Ancient World.* MIT Press, Cambridge.

Sable, Trudy
2001 Mirror, Message, Map, and Mall: Innu Narratives of the Landscape. Paper presented at the Ashkui Conference, Gorsebrook Research Centre, Saint Mary's University, Halifax.

Sabloff, Jeremy A.
1986 Interaction among the Classic Maya polities: a preliminary examination. In *Peer-polity Interaction and Socio-political Change*, edited by Colin Renfrew and John F. Cherry, pp.109-116. Cambridge University Press, Cambridge.

Sanders, William T. and David Webster
1988 The Mesoamerican Urban Tradition. *American Anthropologist* 90:521-546.

Scarborough, Vernon L., and Fred Valdez, Jr.
2009 An Alternative Order: The Dualistic Economies of the Ancient Maya. *Latin American Antiquity* 20:207-227.

Scarborough, Vernon L., and David R. Wilcox (editors)
1991 *The Mesoamerican Ballgame.* University of Arizona Press, Tucson.

Schele, Linda
1991 An Epigraphic History of the Western Maya Region. In *Classic Maya Political History: Archaeological and Hieroglyphic Evidence*, edited by T. Patrick Culbert, pp.72-101. Cambridge University Press, Cambridge.

Schele, Linda, and David A. Freidel
1990 *A Forest of Kings: The Untold Story of the Ancient Maya.* William Morrow and Company, New York.

Schele, Linda, and Julia Guernsey Kappelman
2001 What the Heck's Coatépec? The Enduring Roots of an Enduring Mythology. In *Landscape and Power in Ancient Mesoamerica*, edited by Rex Koontz, Annabeth Headrick, and Kathryn Reese-Taylor, pp. 29-53. Westview Press, Boulder.

Schele, Linda, and Peter Mathews
1991 Royal Visits and Other Intersite Relationships among the Classic Maya. In *Classic Maya Political History: Hieroglyphic and Archaeological Evidence*, edited by T. Patrick Culbert, pp. 226-252. Cambridge University Press, Cambridge.

1998 *The Code of Kings: The Language of Seven Sacred Maya Temples and Tombs.* Scribner, New York.

Schele, Linda, and Mary E. Miller
1986 *The Blood of Kings: Dynasty and Ritual in Maya Art.* Kimball Art Museum, Fort Worth.

Schwake, Sonja
2002 The Ritual Sub-Program of Investigations: The 2002 Research in Group S, Minanhá, Belize. In

Archaeological Investigations in the North Vaca Plateau, Belize: Progress Report of the Fourth (2002) Field Season, edited by Gyles Iannone, Joëlle Chartrand, Rachel Dell, Adam Menzies, Adam Pollock, and Barbara Slim, pp. 115-123. Social Archaeology Research Program, Department of Anthropology, Trent University, Peterborough.

2003 The Ritual Sub-Program of Investigations: The 2003 Research in the Minanha' Periphery. In *Archaeological Investigations in the North Vaca Platuea, Belize:Progress Report of the Fifth (2003) Field Season*, edited by Gyles Iannone and James E. Herbert, pp. 73-82. Social Archaeology Research Program, Department of Anthropology, Trent University, Peterborough.

Seibert, Jeffrey
1999 Excavations in the Ballcourt at Minanhá, Belize. In *Archaeological Investigations in the North Vaca Plateau, Belize: Progress Report of the First (1999) Field Season*, edited by Gyles Iannone, Jeffrey Seibert, and Nadine Gray, pp. 29-41. Social Archaeology Research Program, Department of Anthropology, Trent University, Peterborough.

2002 2002 Excavations in Structure 12A, Minanhá, Belize. In *Archaeological Investigations in the North Vaca Plateau, Belize: Progress Report of the Fourth (2002) Field Season*, edited by Gyles Iannone, Joëlle Chartrand, Rachel Dell, Adam Menzies, Adam Pollock, and Barbara Slim, pp. 7-30. Social Archaeology Research Program, Department of Anthropology, Trent University, Peterborough

Sharer, Robert J.
1994 *The Ancient Maya*, 5th edition. Stanford University Press, Stanford.

Sharer, Robert J., and Loa Traxler
2006 *The Ancient Maya*, 6th edition. Stanford University Press, Stanford.

Shaw, Justine M.
2001 Maya Sacbeob: Form and Function. *Ancient Mesoamerica* 12:261-272.

Smith, Adam T.
1996 Rendering the Political Aesthetic: Political Legitimacy in Urartian Representations of the Built Environment. *Journal of Anthropological Archaeology* 19:131-163.

Smith, Linda Tuhiwai
1999 *Decolonizing Methodologies: Research and Indigenous Peoples*. University of Otago Press and Zed Books, Ltd, Dunedin and London.

Smith, Michael E.
2003 Can We Read Cosmology in Ancient Maya City Plans? Comment on Ashmore and Sabloff. *Latin American Antiquity* 14:221-228.

2004 The Archaeology of Ancient State Economies. *Annual Review of Anthropology* 33:73-102.

2007 Form and Meaning in Earliest Cities: A New Approach to Ancient Urban Planing. *Journal of Planning History* 6:3-47.

2008 Ancient Cities: Do They Hold Lessons for the Modern World? Paper presented at the 73rd Annual Meeting of the Society for American Archaeology, Vancouver, British Columbia.

2009 Just How Comparative is Urban Geography? A Perspective from Archaeology. *Urban Geography* 30:113-117.

Smith, Michael E., and Katharina J. Schreiber
2005 New World States and Empires: Economic and Social Organization. *Journal of Archaeological Research* 13:189-212.

2006 New World States and Empires: Politics, Religion, and Urbanism. *Journal of Archaeological Research* 14:1-52.

Southall, Aiden
1956 *Alur Society*. Oxford University Press, Oxford.

1998 *The City in Time and Space*. Cambridge University Press, Cambridge.

Steward, Julien
1950 *Area Research: Theory and Practice*. Bulletin 63, Social Science Research Council, New York.

1955 [2000] The Patrilineal Band. In *Anthropological Theory: An Introductory History*, 2nd ed., edited by Robert J. McGee and Ron L. Warms, pp. 228-243. Mayfield Publishing Company, Mountain View.

Sugiyama, Saburo
1993 Worldview Materialized in Teotihuacan, Mexico. *Latin American Antiquity* 4:103-29.

Tambiah, Stanley J.
1977 The Galactic Polity: The Structure of Traditional Kingdoms in Southeast Asia. *Annals of the New York Academy of Sciences* 293:69-97.

Taylor, Walter W.
1948 *A Study of Archaeology*. Memoirs of the American Anthropological Association No. 69, Menasha.

Taube, Karl
1993 The Jade Hearth: Centrality, Rulership, and the Classic Maya Temple. In *Lowland Maya Civilisation In The Eighth Century A.D.*, edited by Jeremy A. Sabloff and John S. Henderson, pp. 427-478. Dumbarton Oaks, Washington.

Tilley, Christopher
1994 *A Phenomenology of Landscape: Places, Paths,*

and *Monuments*. Berg, Oxford.

Thomas, Julian
1990 Monuments from the Inside: the Case of the Irish Megalithic Tombs. *World Archaeology* 22:168-178.

2001 Archaeologies of Place and Landscape. In *Archaeological Theory Today*, edited by Ian Hodder, pp. 165-186. Polity, Oxford.

Thompson, J. Eric S.
1931 *Archaeological Investigations in the Southern Cayo District, British Honduras*. Field Museum of Natural History, Chicago.

1954 *The Rise and Fall of Maya Civilisation*. University of Oklahoma Press, Normal.

Tour By Mexico
2003 Pomona. Electronic document. http://www.tourbymexico.com/tabasco/pomona/pomona.htm.jpg , accessed November 14, 2003.

Tozzer, Alfred M.
1907 *A Comparative Study of the Mayas and the Lacandones*. Archaeological Institute of America. Macmillan, New York.

1941 *Landa's Relación de las Cosas de Yucatán*. Papers of the Peabody Museum of American Archaeology and Ethnology 18. Harvard University, Cambridge.

Trigger, Bruce G.
1989 *A History of Archaeological Thought*. Cambridge University Press, Cambridge.

1999 Alternative Archaeologies: Nationalist, Colonialist, Imperialist. In *Contemporary Archaeology in Theory: A Reader*, edited by Robert Preucel and Ian Hodder, pp. 615-631. Blackwell, Oxford.

2003 *Understanding Early Civilisations: A Comparative Study*. Cambridge University Press, New York.

Ucko, Peter, and Robert Layton (editors)
1999 *The Archaeology and Anthropology of Landscape: Shaping Your Landscape*. Routledge, London.

Valdés, Juan Antonio
1997 Tamarandito: Archaeology and Regional Politics in the Petexbatun Region. *Ancient Mesoamerica* 8:321-335.

Vogt, Evon Z.
1969 *Zinacantan: A Maya Community in the Highlands of Chiapas*. Harvard University Press, Cambridge.

1970 *The Zinacantecos of Mexico: A Modern Maya Way of Life*. Holt, Rinehart, and Winston, New York.

Wagner, R.
2000 Condensed Mapping: Myth and the Folding of Space/Space and the Folding of Myth. In *Emplaced Myth: Space, Narrative, and Knowledge in Aboriginal Australia and Papua New Guinea*, edited by A. Rumsey and J. Weiner, pp. 71-78. University of Hawai'i Press, Honolulu.

Walker, William H., and Lisa Lucero
2000 The Depositional History of Ritual and Power. In *Agency in Archaeology*, edited by Marcia Ann-Dobres and John Robb, pp. 130-147. Routledge, London.

Watanabe, John M.
1983 In the World of the Sun: A Cognitive Model of Maya Cosmology. *Man* 18:710-728.

Webster, David
1993 The Study of Maya Warfare: What It Tells Us about the Maya and What It Tells Us about Maya Archaeology. In *Lowland Maya Civilisation In The Eighth Century A.D.*, edited by Jeremy A. Sabloff and J.S. Henderson, pp. 415-444. Dumbarton Oaks, Washington.

Wheatley, Paul
1971 *Pivot of the Four Corners*. University of Chicago Press, Chicago.

1979 *City as Symbol*. University of London Press, London.

White, Leslie
1959 *The Evolution of Culture*. McGraw-Hill, New York.

Willey, Gordon R.
1953 *Prehistoric Settlement Patterns in the Viru Valley, Peru*. Bureau of American Ethnology, Bulletin 155, Washington.

1980 Towards a Holistic View of Ancient Maya Civilization. *Man* 15:249-266.

Willey, Gordon R. (editor)
1956 *Prehistoric Settlement Patterns in the New World*, Viking Fund Publications in Anthropology, No. 23. Wenner-Gren Foundation for Anthropological Research, New York.

Willey, Gordon R., William R. Bullard, Jr., J.B. Glass, and James C. Gifford
1965 *Prehistoric Maya Settlements in the Belize Valley*. Papers of the Peabody Museum of Archaeology and Ethnology, Harvard University, Cambridge.

Willey, Gordon R., A. Ledyard Smith, Gait Toutellot, and Ian Graham
1975 *Excavations at Seibal, Department of the Petén, Guatemala, 1. Introduction: The Site and its Settings*, Memoir 13(1). Peabody Museum of Archaeology and Ethnology, Harvard University, Cambridge.

Yaegar, Jason
2000 The Social Construction of Communities in the

Classic Maya Countryside:Strategies of Affiliation in Western Belize. In *The Archaeology of Communities: A New World Perspective*, edited by Marcello A. Canuto and Jason Yaeger, pp. 123-142. Routledge, London.

Yoffee, Norman
2005 *Myths of the Archaic State: Evolution of the Earliest Cities, States, and Civilizations*. Cambridge University Press, Cambridge.

Yule, George
1996 *The Study of Language*. Cambridge University Press, Cambridge.

www.ingramcontent.com/pod-product-compliance
Lightning Source LLC
Chambersburg PA
CBHW061549010526
44115CB00023B/2989